ASSEMBLY AUTOMATION AND PRODUCT DESIGN

MANUFACTURING ENGINEERING
AND MATERIALS PROCESSING

A Series of Reference Books and Textbooks

SERIES EDITORS

Geoffrey Boothroyd

*Chairman, Department of Industrial
and Manufacturing Engineering
University of Rhode Island
Kingston, Rhode Island*

George E. Dieter

*Dean, College of Engineering
University of Maryland
College Park, Maryland*

OTHER VOLUMES IN PREPARATION

ASSEMBLY AUTOMATION AND PRODUCT DESIGN

GEOFFREY BOOTHROYD
University of Rhode Island
Kingston, Rhode Island

Marcel Dekker, Inc.

New York • Basel • Hong Kong

Library of Congress Cataloging--in--Publication Data

Boothroyd, G. (Geoffrey)
    Assembly automation and product design/Geoffrey Boothroyd.
       p.   cm. -- -- (Manufacturing engineering and materials processing
 ; 37)
    Includes bibliographical references and index.
    ISBN 0-8247-8547-9
    1. Assembly-line methods-- --Automation. 2. Assembling machines.
I. Title. II. Series.
TS178.4.B66   1991
670.42'7-- --dc20   91-25608
                                                CIP

This book is printed on acid-free paper.

MARCEL DEKKER, INC.
270 Madison Avenue, New York, New York 10016

Current printing (last digit):
10 9 8 7 6 5 4 3 2 1

PRINTED IN THE UNITED STATES OF AMERICA

Preface

Portions of this book are based on a book published in 1968 under the title *Mechanized Assembly* by G. Boothroyd and A. H. Redford. In a further edition, entitled *Automatic Assembly* by G. Boothroyd, C. Poli, and L. E. Murch, the original material developed at the University of Salford in England was updated with work carried out at the University of Massachusetts. In those days, it was felt that manufacturing engineers and designers wished to learn about automatic assembly since it appeared to provide a means of improving productivity and competitiveness. Since 1980, however, my colleague, Peter Dewhurst, and I have developed a subject that holds much greater promise for productivity improvement and cost reduction, namely, design for assembly. Our techniques have become widely used and have helped numerous companies introduce new and extremely competitive product designs.

This new text, therefore, includes detailed discussions of design for assembly and, thus, the subject of assembly automation is considered in parallel with that of product design.

Clearly, the first step in considering automation of assembly processes should be careful analysis of the product design for ease of automatic assembly. In addition, analysis of the product for ease of manual assembly should be carried out in order to provide the basis for economic comparisons of automation. Indeed, it is often found that if a product is designed appropriately, manual assembly is so inexpensive that automation cannot be justified. Thus, a whole chapter is devoted to design for manual assembly. Another chapter is devoted to design for high-speed automatic and robot assembly, and a third chapter deals with electronics assembly.

The remaining material has been updated or rewritten as necessary—particularly to reflect interest in the use of general-purpose assembly robots.

The book is intended to appeal to manufacturing and product engineers as well as to engineering students in colleges and universities.

I wish to thank Dr. A. H. Redford for his kind permission to use material published in the original book, *Mechanized Assembly*, and Dr. P. Dewhurst for his contributions to the subject of design for assembly. I also wish to thank Lori Allen and Joanne Pasquazzi for typing the manuscript.

Geoffrey Boothroyd

Contents

4. Automatic Feeding and Orienting—Mechanical Feeders   95

5. Feed Tracks, Escapements, Parts-Placement Mechanisms, and Robots   137

Contents

ASSEMBLY AUTOMATION AND PRODUCT DESIGN

1

Introduction

Since the beginning of the nineteenth century, the increasing need for finished goods in large quantities, especially in the armaments industries, has led engineers to search for and to develop new methods of manufacture or production. As a result of developments in the various manufacturing processes, it is now possible to mass-produce high-quality durable goods at low cost. One of the more important manufacturing processes is the assembly process that is required when two or more component parts are to be secured together.

The history of assembly process development is closely related to the history of the development of mass-production methods. The pioneers of mass production are also the pioneers of modern assembly techniques. Their ideas and concepts have brought significant improvements in the assembly methods employed in high-volume production.

However, although many aspects of manufacturing engineering, especially the parts fabrication processes, have been revolutionized by the application of automation, the technology of the basic assembly process has failed to keep pace. Table 1.1 shows that, in the United States, the percentage of the total labor force involved in the assembly process varies from about 20% for the manufacture of farm machinery to almost 60% for the manufacture of telephone and telegraph equipment. Because of this, assembly costs often account for more than 50% of the total manufacturing costs.

1

Table 1.1 Percentage of Production Workers Involved in Assembly

Industry	Percentage of workers involved in assembly
Motor vehicles	45.6
Aircraft	25.6
Telephone and telegraph	58.9
Farm machinery	20.1
Household refrigerators and freezers	32.0
Typewriters	35.9
Household cooking equipment	38.1
Motorcycles, bicycles, and parts	26.3

Source: 1967 Census of Manufacturers
    U.S. Bureau of the Census

Although, during the last few decades, efforts have been made to reduce assembly costs by the application of high-speed automation and, more recently, by the use of assembly robots, success has been quite limited, and many assembly workers are still using the same basic tools as those employed at the time of the Industrial Revolution.

1.1 HISTORICAL DEVELOPMENT OF THE ASSEMBLY PROCESS

In the early days, the manufacture of the parts and their fitting and assembly were carried out by craftsmen who learned their trade as indentured apprentices. Each part would be tailored to fit its mating parts. Consequently, it was necessary for a craftsman to be an expert in all the various aspects of manufacture and assembly, and training a new craftsman was a long and expensive task. The scale of production was often limited by the availability of trained craftsmen rather than by the demand for the product. This problem was compounded by the reluctance of the craft guilds to increase the number of workers in a particular craft.

The conduct of war, however, requires reliable weapons in large quantities. In 1798, the United States needed a large supply of muskets, and federal arsenals could not meet the demand. Because war with the French was imminent, it was not possible to obtain additional supplies from Europe. Eli Whitney, now recognized as one of the pioneers of mass production, offered to contract to make 10,000 muskets in 28 months. Although it took 10½ years to complete the contract, Whitney's novel ideas on mass production had been successfully proved. At first, Whitney designed templates for each part, but he could not find machinists capable of following the contours. Next, he developed a milling machine that could follow the templates but hand-fitting of the parts was still necessary. Eventually, the factory at New Haven, Connecticut, built especially for the manufacture of the muskets, contained machines for producing interchangeable parts. These machines reduced the skills required by the various workers and allowed significant increases in the rate of production. In a historic demonstration in 1801, Whitney surprised his distinguished visitors when he assembled a musket lock after selecting a set of parts from a random heap.

The results of Eli Whitney's work brought about three primary developments in manufacturing methods. First, parts were manufactured on machines, resulting in consistently higher quality than that of handmade parts. These parts were not interchangeable and, as a consequence, assembly work was simplified. Second, the accuracy of the final product could be maintained at a higher standard; and, third, production rates could be significantly increased. These concepts became known as the American system of manufacture.

Oliver Evans' concept of conveying materials from one place to another without manual effort led eventually to further developments in automation for assembly. In 1793, Evans used three types of conveyors in an automatic flour mill that required only two operators. The first operator poured grain into a hopper, and the second filled sacks with the flour produced by the mill. All the intermediate operations were carried out automatically, with conveyors carrying the material from operation to operation.

A significant contribution to the development of assembly methods was made by Elihu Root. In 1849, Root joined the company that was producing Colt six-shooters. Even though, at that time, the various operations of assembling the component parts were quite simple, he divided these operations into basic units that could be completed more quickly and with less chance of error. Root's division of operations gave rise to the concept "Divide the work and multiply the output." Using this principle, assembly work was reduced to basic operations and, with only short periods of worker training, high efficiencies could be obtained.

Frederick Winslow Taylor was probably the first person to introduce the methods of time and motion study to manufacturing technology. The object was to save the worker's time and energy by making sure that the work and all things associated with the work were placed in the best positions for carrying out the required tasks. Taylor also discovered that any worker has an optimum speed of working which, if exceeded, results in a reduction in overall performance.

Undoubtedly, the principal contributor to the development of modern production and assembly methods was Henry Ford. He described his principles of assembly in the following words:

First, place the tools and then the men in the sequence of the operations so that each part shall travel the least distance whilst in the process of finishing.

Second, use work slides or some other form of carrier so that when a workman completes his operation he drops the part always in the same place which must always be the most convenient place to his hand and if possible have gravity carry the part to the next workman.

Third, use sliding assembly lines by which parts to be assembled are delivered at convenient intervals, spaced to make it easier to work on them.

These principles were gradually applied in the production of the Model T Ford automobile.

The modern assembly-line technique was first employed in the assembly of a flywheel magneto. In the original method, one operator assembled a magneto in 20 min. It was found that when the process was divided into 29 individual operations, carried out by different workers situated at assembly stations spaced along an assembly line, the total assembly time was reduced to 13 min, 10 sec. When the height of the assembly line was raised by 8 in., the time was reduced to 5 min, which was only one fourth of the time required in the original process of assembly. This result encouraged Henry Ford to utilize his system of assembly in other departments of the factory, which were producing subassemblies for the car. Subsequently, this brought a continuous and rapidly increasing flow of subassemblies to those working on the main car assembly. It was found that these workers could not cope with the increased load, and it soon became clear that the main assembly would also have to be carried out on an assembly line. At first, the movement of the main assemblies was achieved simply by pulling them by a rope from station to station. However, even this development produced the amazing result of a reduction in the total time of assembly from 12 hr, 28 min, to 5 hr, 50 min. Eventually, a power-driven endless conveyor was installed; it was flush with the floor and wide enough to accommodate a chassis. Space was provided for workers either to sit or stand while they carried out their operations, and the conveyor moved at a speed of 6 ft/min past 45 separate workstations. With the introduction of this conveyor, the total assembly time was reduced to 93 min. Further improvements led to an even shorter overall assembly time and, eventually, a production rate of one car every 10 sec of the working day was achieved.

Although Ford's target of production had been exceeded, and the overall quality of the product had improved considerably, the assembled products sometimes varied from the precise standards of the hand-built prototypes. Eventually, Ford adopted a method of isolating difficulties and correcting them in advance before actual mass production began. The method was basically to set up a pilot plant, where a complete assembly line was installed, using exactly the same tools, templates, forming devices, gauges, and even the same labor skills that would eventually be used for mass production. This method has now become standard practice for all large assembly plants.

The type of assembly system described above is usually referred to as a manual assembly line, and it is still the most common method of assembling mass- or large-batch-produced products. Since the beginning of the twentieth century, however, methods of replacing manual assembly workers by mechanical devices have been introduced. These devices take the form of automatic assembly devices or workheads with part-feeding mechanisms and, more recently, robots with part trays.

Thus, in the beginning, automated screwdrivers, nut runners, riveters, spot-welding heads, and pick-and-place mechanisms were positioned on transfer devices that moved the assemblies from station to station. Each workhead was supplied with oriented parts, either from a magazine or from an automatic feeding and orienting device—usually a vibratory bowl feeder. The special single-purpose workheads could continually repeat the same operation, usually taking no more than a few seconds. This meant that completed assemblies were produced at rates on the order of 10–30/min. For two-shift working, this translates into an annual production volume of several million.

Automation of this type was usually referred to as mechanization and, because it could be applied only in mass production, its development was closely tied to certain industries: for example, those manufacturing armaments, automobiles, watches and clocks, and other consumer products. Mechanization was used in the manufacture of those individual items such as light bulbs and safety pins that are produced in large quantities. It was probably the process industries, however,—for example, the food, drug, and cosmetic industries— that were the first to apply mechanization on a large scale.

Recent estimates of the proportion of mass-produced durable goods to the total production of durable goods range from 15 to 20%. It is not surprising, therefore, that only about 5% of products are automatically assembled, the remainder being assembled manually. As a result, since World War II, increasing attention has been given to the possibility of using robots in assembly work. It was felt that, because robots are basically versatile and reprogrammable, they could be applied in small- and medium-batch manufacturing situations, which form over 80% of all manufacture.

According to Schwartz [1], George Deval, Jr., patented a programmable transfer device in 1954, which served as the basis for the modern industrial robot. By 1965, several licenses had been issued, and specifications had been outlined for the first modern industrial robot, the Unimate [2]. In 1962, the Unimate Mark 1 prototype was built.

The first uses of industrial robots were in materials handling such as die-casting and punch-press operations and, by 1968, they started to be used in assembly. By 1972, more than 30 different robots were available from 15 manufacturers. Since then, the installation of industrial robots has increased exponentially.

In spite of these developments, the assembly of mechanical and electromechanical assemblies remains difficult to automate except in mass-production quantities. The exception to this is the electronics industry, more specifically printed-circuit-board (PCB) assembly. Because of the special nature of this product, introduced in the early 1950s, it has been found possible to apply assembly automation—even in small-batch production.

At first, components were hand-inserted and their leads hand-soldered. To reduce the soldering time, wave soldering was developed in which all the leads are soldered in one pass of the board through the soldering machine. The next step was automatic insertion of the component leads. Although, initially, most small components were axial-lead components, several large electronics manufacturers developed multiple-head insertion systems. The first company to produce these systems commercially was USM (United Shoe Machinery), now Dynapert Division of Emhart Corporation [2].

The printed-circuit board is an ideal product for the application of assembly automation. It is produced in vast quantities, albeit in a multitude of styles. Assembly of components is carried out in the same direction, and the components types are limited. With the standardization of components now taking place, a high proportion of printed-circuit boards can be assembled entirely automatically. Also, the automatic-insertion machines are easy to program and set up and can perform one to two insertions per second. Consequently, relatively slow manual assembly is often economic for only very small batches.

With the present widespread engineering trend toward replacing mechanical control devices and mechanisms with electronics, PCB assembly automation is now finding broad application; indeed, one of the principal applications of assembly robots is in the insertion of nonstandard (odd-form) electronic components that cannot be handled by the available automatic-insertion machines.

1.2 CHOICE OF ASSEMBLY METHOD

When considering the manufacture of a product, a company must take into account the many factors that affect the choice of assembly method. For a new product, the following considerations are generally important:

1. Suitability of the product design
2. Production rate required
3. Availability of labor
4. Market life of the product

If the product has not been designed with automatic assembly in mind, manual assembly is probably the only possibility. Similarly, automation will not be practical unless the anticipated production rate is high. If labor is plentiful, the degree of automation desirable will depend on the anticipated reduction in cost of assembly and the increase in production rate, assuming the increase can be absorbed by the market. The capital investment in automatic machinery must usually be amortized over the market life of the product. Clearly, if the market life of the product is short, automation will be difficult to justify.

A shortage of assembly workers will often lead a manufacturer to consider automatic assembly when manual assembly would be cheaper. This situation frequently arises when a rapid increase in demand for a product occurs. Another reason for considering automation in a situation in which manual assembly would be more economical is for research and development purposes, where experience in the application of new equipment and techniques is considered desirable. Many of the early applications of assembly robots were conducted on this basis.

Following are some of the advantages of automation applied in appropriate circumstances:

1. Increased productivity and reduction in costs
2. A more consistent product with higher reliability
3. Removal of operators from hazardous operations
4. The opportunity to reconsider the design of the product

Productivity is the relationship between the output of goods and services and one or more of the inputs—labor, capital, goods, and natural resources. It is expressed as a ratio of output divided by input. Both output and input can be measured in different ways, none of them being satisfactory for all purposes. The most common way of defining and measuring productivity is output per man-hour; usually referred to as labor productivity. This measure of productivity is easy to understand, and it is the only measure for which reliable data have been accumulated over the years.

However, a more realistic way of defining productivity is the ratio of output divided by total input, usually referred to as total productivity. Total productivity is difficult to measure because it is not generally agreed how the various contributions of labor, machinery, capital, etc., should be weighed relative to each other. Also, it is possible to increase labor productivity while at the same time reducing total productivity. To take a hypothetical example, suppose a company is persuaded to install a machine that costs $100,000 and that effectively does a job that is the equivalent of one worker. The effect will be to raise the output per man-hour (increase labor productivity) because fewer workers will be needed to maintain the same output of manufactured units. However, since it is unlikely to be economical for a company to spend much more than $100,000 to replace one worker, the capital investment is not worthwhile, and it results in lower total productivity and increased costs. In the long run, however, economic considerations will be taken into account when investment in equipment or labor is made, and improvements in labor productivity will generally be accompanied by corresponding improvements in total productivity.

Productivity in manufacturing should be considered particularly important because of the relatively high impact that this sector of industry has on the generation of national wealth. In the United States, manufacturing absorbs over 25% of the nation's work force and is the most significant single item in the national income accounts, contributing about 30% of the gross national product. This figure is about nine times the contribution of agriculture and construction and three times that of finance and insurance. In 1974, manufacturing industries accounted for over two thirds of the wealth-producing activities of the United States [3]. Within manufacturing in the United States, the discrete-parts/durable goods industries clearly form a major target for productivity improvements since these are under direct attack from imports of economically priced high-quality items. These industries manufacture engines; farm, metal-working, and electrical machinery and equipment; home electronic and electrical equipment; communications equipment; motor vehicles; aircraft; ships; and photographic equipment. This important group of industries contribute approximately 13% of the GNP, and their output is 46% of that of the manufacturing sectors and 80% of that of the durable-goods manufacturing industries. These industries are highly significant in international trade, their output constituting 80% of total manufacturing exports. However, these industries are not generally the highly efficient, highly automated mass-production units that one is often led to believe they are. The great bulk of the products of these industries are produced in small to medium batches in inefficient factories using relatively ancient machines and tools. These industries typically depend on manual labor for the handling and assembly of parts; labor provided with tools no more sophisticated than screwdrivers, wrenches, and hammers. It is not surprising, therefore, that, as we have seen, for a wide variety of manufacturing industries, assembly accounts for more than 50% of the total manufacturing cost of a product and more than 40% of the labor force. This means that assembly should be given high priority in attempts to improve manufacturing productivity.

In the past, in most manufacturing industries, when a new product was considered, careful thought was given to how the product would function, to its appearance and, sometimes, to its reliability. However, little thought was given to how easily the product could be assembled and how easily the various parts could be manufactured. This philosophy is often referred to as the "over-the-wall" approach, or "we design it—you make it." In other words, there is an imaginary wall between the design and manufacturing functions; designs are thrown "over the wall" to manufacturing, as illustrated in Fig. 1.1. This attitude is particularly serious as it affects assembly. The fundamental reason for this is that most part manufacturing is accomplished on machines performing tasks that physically cannot be performed manually, whereas machines that can perform even a small fraction of the selection, inspection, and manipulative capabilities of a manual assembly worker are rare. This has resulted in great reliance on the versatility of assembly workers, particularly in the design of a product.

For example, an assembly worker can quickly conduct a visual inspection of the part to be assembled, and can discard obviously defective parts, whereas elaborate inspection systems would often be required to detect even the most obviously defective part. If an attempt is made to assemble a part that appears to be acceptable but is in fact defective, an assembly worker, after unsuccessfully trying to complete the assembly, can reject the part quickly without a significant loss in production. In automatic assembly, however, the part might

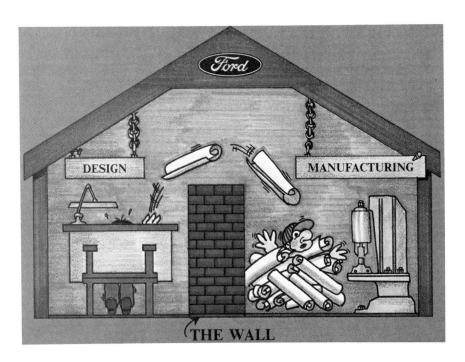

Fig. 1.1 Illustrating "over the wall" design (from Ref. 4).

cause a stoppage of an automatic workhead or robot, resulting in system down-time while the fault is located and corrected. On the other hand, if a part has only a minor defect, an assembly worker may be able to complete the assembly, but the resulting product may not be satisfactory. It is often suggested that one advantage of automatic assembly is that it ensures a product of consistently high quality because the machine cannot handle parts that do not conform to the required specifications. Another advantage is that automatic assembly forces ease of assembly to be considered in the design of the product.

In some situations, assembly by manual workers would be hazardous because of high temperatures and the presence of toxic, or even explosive, substances. Under these circumstances, productivity and cost considerations become less important.

1.3 SOCIAL EFFECTS OF AUTOMATION

Much has been said and written regarding the impact of automation and robots in industry. Newspapers and television have left us with the impression that all consumer products will soon be assembled by general-purpose robots. Nothing could be further from the truth. Often, publicity such as this leads many an industrial manager to inquire why their own company is not using robots in this way and to issue directives to investigate the possibility. An assembly robot is then purchased, and suitable applications are sought. This turns out to be surprisingly difficult, and what usually follows is a full-scale development of a robot assembly system so that the various problem areas can be uncovered. The system thus developed was never meant to be economic although that is not always admitted. In fact, assembly systems based on a single general-purpose assembly robot that performs all the necessary assembly operations are difficult to justify on economic grounds. The central reason for this is that the peripheral equipment (feeders, grippers) needed to build an economic robot assembly station has not yet been developed. The practical difficulties are severe, so that there is no justification for prophesying mass unemployment as a result of the introduction of assembly robots. Moreover, history has shown that special-purpose one-of-a-kind assembly automation (which is relatively easy although expensive to apply) has not had the kind of impact that was feared 25 years ago. In some limited areas, such as the spot-welding of car bodies, industrial robots have made an impact. Special-purpose robots (or programmable automatic insertion machines) are now used for 50% of printed-circuit-board assembly. However, the application of general-purpose robots in batch assembly is, like all other technological changes, taking place slowly. It should be understood that industrial robots are simply one more tool in the techniques available to manufacturing engineers for improving productivity in manufacturing. Considering the overall picture, the robot is not proving to be a particularly effective tool in assembly. Indeed, much greater improvements in manufacturing productivity can be obtained by carefully considering ease of assembly during the design of the products.

It is appropriate to address more carefully those fears that robots are going to have serious adverse effects on employment in manufacturing. The following quotation is taken from the evidence of an important industrialist addressing a U.S. Senate subcommittee on labor and public welfare [5]:

> From a technological point of view automation is working; but the same cannot be said so confidently from the human point of view. The technologists have done and are doing their job. They have developed and are developing equipment that works miracles. But as is too often the case in this age of the widening gap between scientific progress and man's ability to cope with it, we have failed to keep pace.
>
> Much of this failure is due, I think, to the existence of a number of myths about automation, ... The most seductive of these is the claim that, for a number of reasons, automation is not going to eliminate many jobs .... Personally, I think... that automation is a major factor in eliminating jobs in the United States at the rate of more than 40,000 per week, as previous estimates have put it.

These observations are quoted from Senate hearings in 1963—28 years ago! Even before that, in 1950, the famous M.I.T. professor of mathematics Norbert Wiener stated:

> Let us remember that the automatic machine... is the precise economic equivalent of slave labor. Any labor which competes with slave labor must accept the economic conditions of slave labor. It is perfectly clear that this will produce an unemployment situation, in comparison with which... the depression of the thirties will seem a pleasant joke.

It is amusing to look back in retrospect at serious predictions made by famous and influential individuals and see just how wrong they were. However, the problem is that equally famous people are currently making similar pronouncements about the automatic factory, which was considered to be just around the corner 20 years ago and is still just around the corner! It is worth examining these alarmist views a little more carefully because of the very real and adverse effects they can have on public opinion. These views are generally based on two false premises:

1. That the introduction of improved techniques for the manufacture of goods produces rapid and significant changes in productivity;

2. that improvements in productivity have an overall negative effect on employment.

History shows that the introduction of improved manufacturing techniques takes place very slowly. With specific reference to assembly robots, an M.I.T. professor puts it as follows [6]:

There has been in recent years a great deal of publicity associated with robotics. The implication has been that great progress is being made in implementing robot technology to perform assembly tasks. In fact progress during the last ten years has been slow and steady. Present perception in the popular press is that robots are about to take over many manufacturing tasks. Yet there is a growing awareness that this is not so. The rate of progress in this area is accelerating as more money and more interest are being directed toward the problems. The automated factory of the future is still many years away and steps in that direction are being taken at a pace which will allow us, if we so choose, to study and make enlightened decisions about the effects of implementation of flexible automation on unemployment, quality and structure of the work environment, and quality of the workpiece produced.

Even though this statement is reasonable and considered, there is the implication in the last sentence that automation will have the effect of increasing unemployment. Regarding this common premise, it has long been established [7] that there is little, if any, correlation between productivity changes and changes in employment. Recent employment problems in the U.S. auto industry, for example, have arisen from the lack of manufacturing productivity improvement rather than the opposite. Certainly there is no evidence that manufacturing process innovation is, on balance, adverse to employment.

More recently, in summarizing a study of technology and employment [8], Cyert and Mowery stated:

(i) Historically, technological change and productivity growth have been associated with expanding rather than contracting total employment and rising earnings. The future will see little change in this pattern. As in the past, however, there will be declines in specific industries and growth in others, and some individuals will be displaced. Technological change in the U.S. economy is not the sole or even the most important cause of these dislocations.

(ii) The adoption of new technologies generally is gradual rather than sudden. The employment impacts of new technologies are realized through the diffusion and adoption of technology, which typically take a considerable amount of time. The employment impacts of new technologies therefore are likely to be felt more gradually than the employment impacts of other factors, such as changes in exchange rates. The gradual pace of technological change should simplify somewhat the development and implementation of adjustment policies to help affected workers.

(iii) Within today's international economic environment, slow adoption by U.S. firms (relative to other industrial nations) of productivity-increasing technologies is likely to cause more job displacement than the rapid adoption of such technologies. Much of the job displacement since 1980 does not reflect a sudden increase in the adoption of labor saving innovations but instead is due in part to increased U.S. imports and sluggish exports, which in turn reflect macroeconomic forces (the large U.S. budget deficit and the high foreign exchange value of the dollar during 1980–1985), slow adoption of some technologies in U.S. manufacturing, and other factors.

(iv) The rate of technology transfer increasingly incorporates significant research findings and innovations. In many technologies, the United States no longer commands a significant lead over industrial competitor nations. Moreover, technology "gaps" (the time it takes another country to become competitive with U.S. industry or for U.S. firms to absorb foreign technologies) are likely to be shorter in the future.

In conclusion, it can be said that justification for the use of assembly automation equipment can be made on economic grounds (which is quite difficult to do) or because the supply of local manual labor becomes inadequate to meet the demand. In the past, the latter has most often been the real justification— without completely disregarding the first, of course. Thus, the real social impact of the use of robots in assembly is unlikely to be of major proportions. However, the imaginary social impacts are clearly not insignificant, and efforts should be made to obtain realistic data and predictions for the future and to use this to advance the state of the art and to allay fears of widespread unemployment.

Turning to the effects of product design, it can be stated that improvements in product design leading to greater economy in the manufacture of parts and the assembly of products will always lead to improvements in both labor and total productivity. To design a product for ease of assembly requires no expenditure on capital equipment, and yet the significant reductions in assembly times have a marked effect on productivity.

In fact, the design of products for ease of assembly has much greater potential for reducing costs and improving productivity than assembly automation. This is illustrated by the example shown in Fig. 1.2. This graph shows clearly that automation becomes less attractive as the product design is improved. For the original design manufactured in large volumes, high-speed assembly automation would give an 86% reduction in assembly costs and, for medium

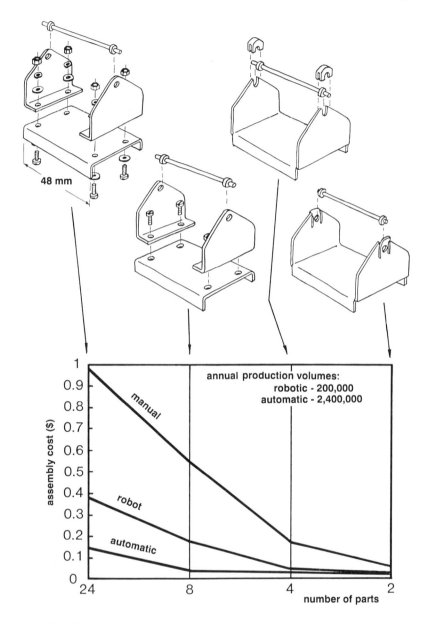

Fig. 1.2 Example of the effect of design for assembly (from Ref. 9).

production volumes, robot assembly would give a 61% reduction. However, with the most efficient design consisting of only two parts, DFA (design for assembly) gives a 92% reduction in manual assembly costs and, for this design, the further benefits obtained through automation are negligible.

In the following chapters, the basic components of assembly machines are presented, and the overall performance of assembly systems is discussed. Finally, detailed analyses of the suitability of parts and products for automatic assembly are presented.

REFERENCES

1. Schwartz, W.H., "An Assembly Hall of Fame," *Assembly Engineering*, Jan. 1988, p.30.
2. Unimation, Inc., Danbury, Conn.
3. "The National Role and Importance of Manufacturing Engineering and Advanced Manufacturing Technology," Position Paper of the S.M.E. (Society of Manufacturing Engineers), May 8, 1978.
4. Munro, S., Illustrating "over the wall" design. Private communication.
5. Terborgh, G., "The Automation Hysteria," Norton, New York, 1965.
6. Seering, W.P., and Gordon, S.J., "Review of Literature on Automated Assembly," Dept. of Mechanical Engineering, MIT, Cambridge, MA., Nov. 1983.
7. Aron, P., "The Robot Scene in Japan: An Update," Report No. 26, Diawan Securities American Inc., New York, 1983.
8. Cyert, R.M., and Mowery, D.C., "Technology and Employment," National Academy Press, Washington, D.C., 1987
9. Boothroyd, G., "Design for Assembly—The Key to Design for Manufacture," *International Journal of Advanced Manufacturing Technology*, Vol. 2, No. 3, 1987.

2

Automatic Assembly Transfer Systems

In automatic assembly, the various individual assembly operations are generally carried out at separate workstations. For this method of assembly, a machine is required for transferring the partly completed assemblies from workstation to workstation, and a means must be provided to ensure that no relative motion exists between the assembly and the workhead or robot while the operation is being carried out. As the assembly passes from station to station, it is necessary that it be maintained in the required attitude. For this purpose, the assembly is usually built up on a base or work carrier, and the machine is designed to transfer the work carrier from station to station; an example of a typical work carrier is shown in Fig. 2.1. Assembly machines are usually classified according to the system adopted for transferring the work carriers (Fig. 2.2). Thus, an in-line assembly machine is one in which the work carriers are transferred in-line along a straight slideway, and a rotary machine is one in which the work carriers move in a circular path. In both types of machine, the transfer of work carriers may be continuous or intermittent.

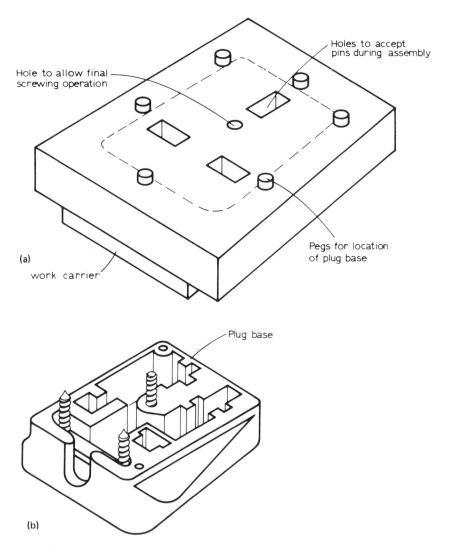

Holes to accept
pins during assembly

Hole to allow final
screwing operation

Pegs for location
of plug base

(a)

work carrier

Plug base

(b)

Fig. 2.1 Work carrier suitable for holding and transferring three-pin power plug base.

2.1 CONTINUOUS TRANSFER

With continuous transfer, the work carriers are moving at constant speed while the workheads index back and forth. Alternatively, the workheads move in a circular path tangential to the motion of the work carriers. In either case, the

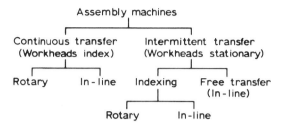

Fig. 2.2 Basic types of assembly machines.

assembly operations are carried out during the period in which the workheads are keeping pace with the work carriers.

Continuous-transfer systems have limited application in automatic assembly because the workheads and associated equipment are often heavy and must therefore remain stationary. It is also difficult to maintain sufficiently accurate alignment between the workheads and work carriers during the operation cycle since both are moving. Continuous-transfer machines are most common in industries such as food processing or cosmetics, where bottles and jars are being filled with liquids.

2.2 INTERMITTENT TRANSFER

Intermittent transfer is the more common system employed for automatic assembly. As the name implies, the work carriers are transferred intermittently, and the workheads remain stationary. Often, the transfer of all the work carriers occurs simultaneously, and the carriers then remain stationary to allow time for the assembly operations. These machines may be termed indexing machines, and typical examples of the rotary and in-line types of indexing machines are shown in Figs. 2.3 and 2.4, respectively. With rotary indexing machines, indexing of the table brings the work carriers under the various workheads in turn, and assembly of the product is completed during one revolution of the table. Thus, at the appropriate station, a completed product may be taken from the machine after each index. The in-line indexing machine works on a similar principle but, in this case, a completed product is removed from the end of the line after each index. With in-line machines, provision must be made for returning the empty work carriers to the beginning of the line. The transfer mechanism on in-line machines is generally one of two types: the shunting work carrier or the belt-driven work carrier.

The shunting work-carrier transfer system is shown in Fig. 2.5. In this system, the work carriers have lengths equal to the distance moved during one index. Positions are available for work carriers at the beginning and end of the

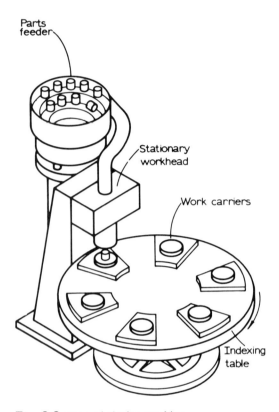

Fig. 2.3 Rotary indexing machine.

assembly line, where no assembly takes place. At the start of the cycle of operations, the work carrier position at the end of the line is vacant. A mechanism pushes the line of work carriers up to a stop at the end of the line, and this indexes the work carriers one position. The piston then withdraws, and the completed assembly at the end of the line is removed. The empty work carrier from a previous cycle that has been delivered by the return conveyor is raised into position at the beginning of the assembly line.

Although the system described here operates in the vertical plane, the return of work carriers can also be accomplished in the horizontal plane. In this case, transfer from the assembly line to the return conveyor (and vice versa) is simpler, but greater floor area is used. In practice, when operating in the horizontal plane, it is more usual to dispense with the rapid return conveyor and to fit further assembly heads and associated transfer equipment in its

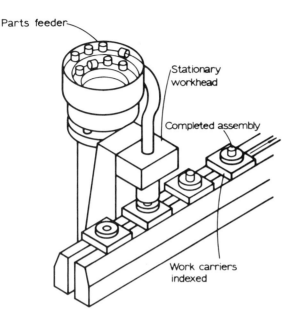

Fig. 2.4 In-line indexing machine.

place (Fig. 2.6). This system has the disadvantage that access to the various workheads may be difficult.

A further disadvantage with all shunting work-carrier systems is that the work carriers themselves must be accurately manufactured. For example, if an error of 0.025 mm were to occur on the length of each work carrier in a 20-station machine, an error in alignment of 0.50 mm would occur at the last station. This error could create serious difficulties in the operation of the workheads. However, in all in-line transfer machines, it is usual for each work carrier, after transfer, to be finally positioned and locked by a locating plunger before the assembly operation is initiated.

The belt-driven work-carrier transfer system is illustrated in Fig. 2.7. Basically, this machine uses an indexing mechanism that drives a belt or flexible steel band to which the work carriers are attached. The work carriers are spaced to correspond to the distance between the workheads.

Instead of attaching the work carriers rigidly to the belt, it is possible to employ a chain that has attachments to push the work carriers along guides. In this case, the chain index can be arranged to leave the work carriers short of their final position, allowing location plungers to bring them into line with the workheads.

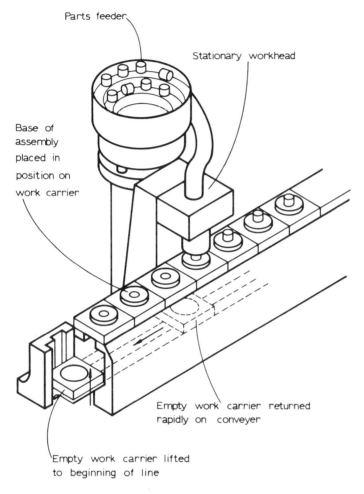

Parts feeder

Stationary workhead

Base of
assembly
placed in
position on
work carrier

Empty work carrier returned
rapidly on conveyer

Empty work carrier lifted
to beginning of line

Fig. 2.5 In-line transfer machine with shunting work carriers returned in vertical plane.

2.3 INDEXING MECHANISMS

Huby [1] lists the factors affecting the choice of indexing mechanism for an assembly machine as follows:

1.   The required life of the machine
2.   The dynamic torque capacity required
3.   The static torque capacity

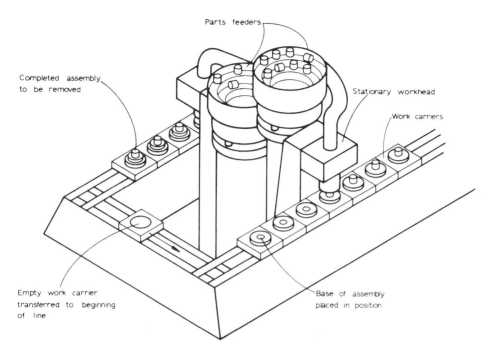

Fig. 2.6 In-line transfer machine with shunting work carriers returned in horizontal plane.

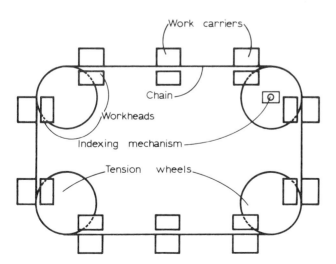

Fig. 2.7 Belt-driven transfer system.

4. The power souce required to drive the mechanism
5. The acceleration pattern required
6. The accuracy of positioning required from the indexing unit

Generally, an increase in the size of a mechanism increases its life. Experience shows which mechanisms usually give longest life for given applications; this is discussed later.

The dynamic torque capacity is the torque that must be supplied by the indexing unit during the index of a fully loaded machine. The dynamic torque capacity is found by adding the effects of inertia and friction and multiplying by the life factor of the unit, the latter factor being found from experience with use of the indexing units.

The static torque capacity is the sum of the torques produced at the unit by the operation of the workheads. If individual location plungers are employed at each workhead, these plungers are usually designed to withstand the forces applied by the workheads; in such a case, the static torque capacity required from the indexing unit will probably be negligible. The power required to drive an indexing unit is obtained from the dynamic torque applied to the unit during the machine index.

The form of the acceleration curve for the indexing unit may be very important when there is any possibility that a partially completed assembly may be disturbed during the machine index. A smooth acceleration curve will also reduce the peak dynamic torque and will thus assist the driving motor in maintaining a reasonably constant speed during indexing, thereby increasing the life of the machine. The accuracy of indexing required will not be great if locating plungers are employed to perform the final location of the work carriers or indexing table.

Various indexing mechanisms are available for use on automatic assembly machines; typical examples are given in Figs. 2.8–2.10. These mechanisms fall into two principal categories: those that convert intermittent translational motion (usually provided by a piston) into angular motion by means of a rack and pinion or ratchet and pawl (Fig. 2.8); and those that are continuously driven, such as the Geneva mechanism (Fig. 2.9) or the crossover or scroll cam shown in Fig. 2.10.

For all but very low-speed indexing or very small indexing tables, the rack-and-pinion or ratchet-and-pawl mechanisms are unsuitable because they have a tendency to overshoot. The acceleration properties of both these systems are governed entirely by the acceleration pattern of the linear power source. To ensure a fairly constant indexing time, if the power source is a pneumatic cylinder, it is usual to underload the cylinder, in which case the accelerations at the beginning and end of the stroke are very high and produce undesirable shocks. The ratchet and pawl mechanism requires a takeup move-

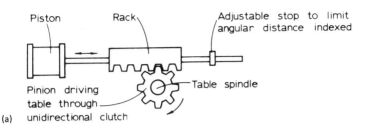

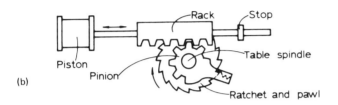

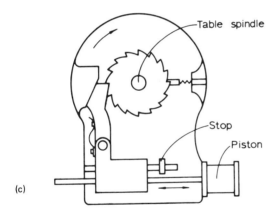

Fig. 2.8 Indexing mechanisms: (a) rack and pinion with unidirectional clutch; (b) rack and pinion with ratchet and pawl; (c) ratchet and pawl.

ment and must be fairly robust if it is to have a long life. The weakest point in the mechanism is usually the pawl pin and, if this is not well lubricated, the pawl will stick and indexing will not occur.

The Geneva-type indexing mechanism has more general application in assembly machines, but its cost is higher than the mechanisms described earlier. It is capable of transmitting a high torque relative to its size and has a smooth acceleration curve. However, it has a high peak dynamic torque

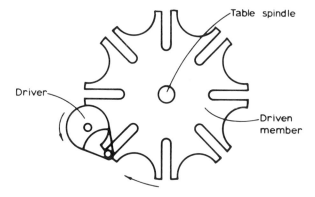

Fig. 2.9 Geneva mechanism.

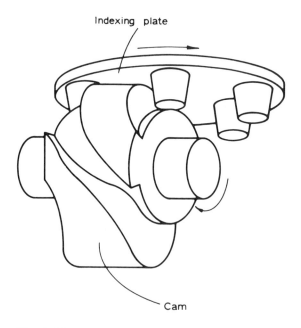

Fig. 2.10 Crossover cam indexing unit.

immediately before and after the reversal from positive to negative acceleration. In its basic form, the Geneva mechanism has a fairly short life, but wear can be compensated for by adjustment of the centers. The weakest point in the mechanism is the indexing pin, but breakages of this part can be averted by careful design and avoidance of undue shock reactions from the assembly machine. A characteristic of the Geneva mechanism is its restriction on the number of stops per revolution. This is due primarily to the accelerations that occur with three-stop and more than eight-stop mechanisms.

In a Geneva mechanism, the smaller the number of stops, the greater the adverse mechanical advantage between the driver and the driven members. This results in a high indexing velocity at the center of the indexing movement and gives a very peaked acceleration graph. On a three-stop Geneva, this peaking becomes very pronounced and, since the mechanical advantage is very high at the center of the movement, the torque applied to the index plate is greatly reduced when it is most required. The solution to these problems results in very large mechanisms relative to the output torque available.

As the number of stops provided by a Geneva mechanism increases, the initial and final accelerations during indexing increase although the peak torque is reduced. This is due to the increased difficulty of placing the driver center close to the tangent of the indexing slot on the driven member.

For a unit running in an oil bath, the clearance between the driver and driven members during the locking movement is approximately 0.025 mm. To allow for wear in this region, it is usual to provide a small center-distance adjustment between the two members. The clearance established after adjustments is the main factor governing the indexing accuracy of the unit, and this will generally become less accurate as the number of stops is increased. Because of the limitations in accuracy, it is usual to employ a Geneva mechanism in conjunction with a location plunger; in this case, a relatively cheap and accurate method of indexing is obtained.

The crossover-cam type of indexing mechanism shown in Fig. 2.10 is capable of transmitting a high torque, has a good acceleration characteristic, and is probably the most consistent and accurate form of indexing mechanism. Its cost is higher than that of the alternative mechanisms described earlier, and it has the minor disadvantage of being rather bulky. The acceleration characteristics are not fixed as with other types of indexing mechanisms, but a crossover cam can be designed to give almost any required form of acceleration curve. The normal type of cam is designed to transmit a modified trapezoidal form of acceleration curve, resulting in a low peak dynamic torque and fairly low mean torque. The cam can be designed to give a wide range of stops per revolution of the index plate, and the indexing is inherently accurate. A further advantage is that it always has at least two indexing pins in contact with the cam.

Figure 2.11 shows the acceleration patterns of the modified trapezoid, sine, and modified sine cams and the Geneva mechanism for the complete index of a four-stop unit. It can be seen that the modified trapezoidal form gives the best pattern for the smoothest operation and lowest peaking. The sine and modified sine both give smooth acceleration, but the peak torque is increased whereas, with the Geneva mechanism, the slight initial shock loading and the peaking at the reversal of the acceleration are clearly evident.

2.4 OPERATOR-PACED FREE-TRANSFER MACHINE

With all the transfer systems described earlier, it is usual for the cycle of operations to occur at a fixed rate, and any manual operations involved must keep pace; this is referred to as *machine pacing*. Machines are available, however, for which a new cycle of operations can be initiated only when signals are received, indicating that all the previous operations have been completed. This is referred to as *operator pacing*.

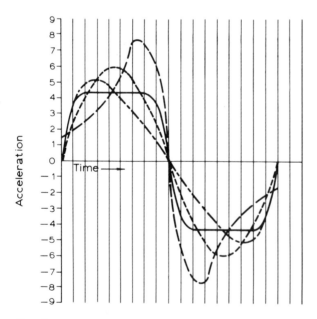

Fig. 2.11 Comparison of acceleration curves for a Geneva mechanism and various designs of crossover cams: modified trapezoidal, — ; four-stop Geneva, — — ; modified sine, — – — – ; sine, ---- . (Adapted from Ref. 1).

One basic characteristic common to all the systems described is that a breakdown of any individual workhead will stop the whole machine, and production will cease until the fault has been cleared. One type of in-line intermittent operator-paced machine, known as a *free-transfer machine* (Fig. 2.12), does not have this limitation. In this design, the spacing of the workstations is such that buffer stocks of assemblies can accumulate between adjacent stations. Each workhead or operator works independently, and the assembly process is initiated by the arrival of a work carrier at the station. The first operation is to lift the work carrier clear of the conveyor and clamp it in position. After the assembly operation has been completed, the work carrier is released and transferred to the next station by the conveyor, provided that a vacant space is available. Thus, on a free-transfer machine, a fault at any one station will not necessarily prevent the other stations from working. It will be seen later that this can be an important factor when considering the economics of various transfer machines for automatic assembly.

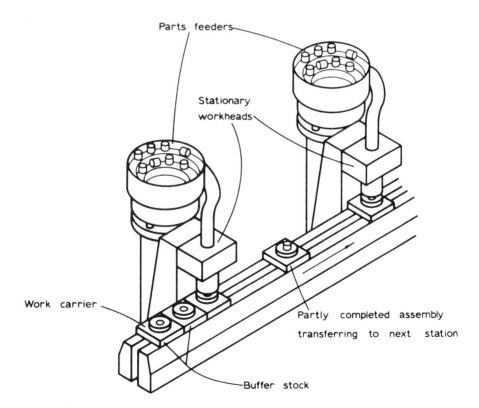

Fig. 2.12 In-line free-transfer machine.

REFERENCE

1. E. Huby, "Assembly Machine Transfer Systems," paper presented at the Conference on Mechanized Assembly, July 1966, Royal College of Advanced Technology, Salford, England.

3

Automatic Feeding and Orienting —
Vibratory Feeders

The vibratory-bowl feeder is the most versatile of all hopper feeding devices for small engineering parts. In this feeder (Fig. 3.1), the track along which the parts travel is helical and passes around the inside wall of a shallow cylindrical hopper or bowl. The bowl is usually supported on three or four sets of inclined leaf springs secured to a heavy base. Vibration is applied to the bowl from an electromagnet mounted on the base, and the support system constrains the movement of the bowl so that it has a torsional vibration about its vertical axis, coupled with a linear vertical vibration. The motion is such that any small portion of the inclined track vibrates along a short, approximately straight path, which is inclined to the horizontal at an angle greater than that of the track. When component parts are placed in the bowl, the effect of the vibratory motion is to cause them to climb up the track to the outlet at the top of the bowl. Before considering the characteristics of vibratory-bowl feeders, it is necessary to examine the mechanics of vibratory conveying. For this purpose, it is convenient to deal with the motion of a part on a straight vibrating track that is inclined at a small angle to the horizontal.

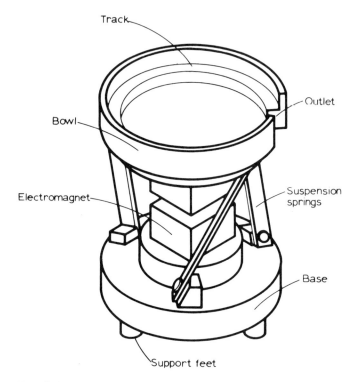

Fig. 3.1 Vibratory-bowl feeder.

3.1 MECHANICS OF VIBRATORY CONVEYING

In the following analysis, the track of a vibratory feeder is assumed to move bodily with simple harmonic motion along a straight path inclined at an angle $[\theta + \psi]$ to the horizontal, as shown in Fig. 3.2. The angle of inclination of the track is θ, and ψ is the angle between the track and its line of vibration. The frequency of vibration f (usually 60 Hz, in practice) is conveniently expressed in this analysis as $\omega = 2\pi f$ rad, where ω is the angular frequency of vibration. The amplitude of vibration a_0 and the instantaneous velocity and acceleration of the track may all be resolved in directions parallel and normal to the track. These components will be referred to as *parallel and normal motions* and will be indicated by the suffices p and n, respectively.

It is assumed in the analysis that the motion of a part of mass m_p is independent of its shape and that air resistance is negligible. It is also assumed that there is no tendency for the part to roll down the track.

It is useful to consider the behavior of a part that is placed on a track whose amplitude of vibration is increased gradually from zero. For small amplitudes,

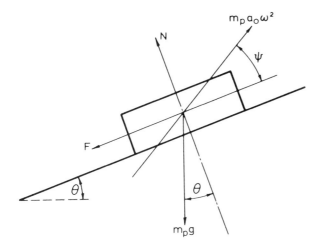

Fig. 3.2 Force acting on a part in vibratory feeding.

the part will remain stationary on the track because the parallel inertia force acting on the part will be too small to overcome the frictional resistance F between the part and the track. Figure 3.2 shows the maximum inertia force acting on the part when the track is at the upper limit of its motion. This force has parallel and normal components of $m_p a_0 \omega^2 \cos \psi$ and $m_p a_0 \omega^2 \sin \psi$, respectively, and it can be seen that, for sliding up the track to occur,

$$m_p a_0 \omega^2 \cos \psi > m_p g \sin \theta + F \qquad (3.1)$$

where

$$F = \mu_s N = \mu_s [m_p g \cos \theta - m_p a_0 \omega^2 \sin \psi] \qquad (3.2)$$

and where μ_s is the coefficient of static friction between the part and the track. The condition for forward sliding up the track to occur is, therefore, given by combining Eqs. (3.1) and (3.2). Thus,

$$\frac{a_0 \omega^2}{g} > \frac{\mu_s \cos \theta + \sin \theta}{\cos \psi + \mu_s \sin \psi} \qquad (3.3)$$

Similarly, it can be shown that, for backward sliding to occur during the vibration cycle,

$$\frac{a_0 \omega^2}{g} > \frac{\mu_s \cos \theta - \sin \theta}{\cos \psi - \mu_s \sin \psi} \qquad (3.4)$$

The operating conditions of a vibratory conveyor may be expressed in terms of the dimensionless normal track acceleration A_n/g_n, where A_n is the normal track acceleration $(A_n = a_n \omega^2 = a_0 \omega^2 \sin \psi)$, g_n the normal acceleration due to gravity $(= g \cos \theta)$, and g the acceleration due to gravity $(= 9.81$ m/s$^2)$. Thus,

$$\frac{A_n}{g_n} = \frac{a_0 \omega^2 \sin \psi}{g \cos \theta} \tag{3.5}$$

Substitution of Eq. (3.5) in Eqs. (3.3) and (3.4) gives, for forward sliding,

$$\frac{A_n}{g_n} > \frac{\mu_s + \tan \theta}{\cot \psi + \mu_s} \tag{3.6}$$

and for backward sliding,

$$\frac{A_n}{g_n} > \frac{\mu_s - \tan \theta}{\cot \psi - \mu_s} \tag{3.7}$$

For values of $\mu_s = 0.8$, $\theta = 3$ deg (0.95 rad) and $\psi = 30$ deg (0.52 rad), Eqs. (3.6) and (3.7) show that the ratio A_n/g_n must be greater than 0.34 for forward sliding to occur and greater than 0.8 for backward sliding. With these conditions, it is clear that, for all amplitudes of vibration giving a value of A_n/g_n greater than 0.34, forward sliding will predominate and the part will climb the track, sliding forward or both forward and backward during each vibration cycle.

The limiting condition for forward conveying to occur is given by comparing Eqs. (3.6) and (3.7). Thus, for forward conveying,

$$\tan \psi > \frac{\tan \theta}{\mu_s^2}$$

or, when θ is small,

$$\tan \psi > \frac{\theta}{\mu_s^2} \tag{3.8}$$

For values of $\mu_s = 0.8$ and $\theta = 3$ deg (0.05), ψ must be greater than 4.7 deg (0.08 rad) for forward conveying to occur.

For sufficiently large vibration amplitudes, the part will leave the track and "hop" forward during each cycle. This can occur only when the normal reaction N between the part and the track becomes zero. From Fig. 3.2,

$$N = m_p g \cos \theta - m_p a_0 \omega^2 \sin \psi \tag{3.9}$$

and, therefore, for the part to leave the track,

$$\frac{a_0 \omega^2}{g} > \frac{\cos \theta}{\sin \psi}$$

or

$$\frac{A_n}{g_n} > 1.0 \qquad (3.10)$$

It is clear from the earlier examples, however, that the part slides forward before it leaves the track during each cycle. Figure 3.3 graphically illustrates these equations, showing the effect of the vibration angle ψ on the limiting values of the dimensionless normal acceleration A_n/g_n for forward sliding to occur, for both forward and backward sliding to occur, and for the part to hop along the track.

The detailed types of motion that may occur in vibratory feeding have been described in the literature [1]. For all conditions, the part starts to slide forward at some instant when the track is nearing the upper limit of its motion. When there is no hopping mode, this forward sliding continues until the track is nearing the lower limit of its motion, at which point the part may remain stationary relative to the track or slide backward until the cycle is complete. In some cases, the stationary period is followed by a period of backward sliding only or of backward sliding followed by yet another stationary period. Finally, the forward sliding is followed by a period of backward sliding and then a stationary period to complete the cycle.

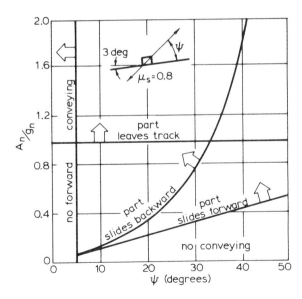

Fig. 3.3 Limiting conditions for various modes of vibratory conveying. A_n is the normal track acceleration and g_n the normal gravitational acceleration.

Analysis and experiment have shown that higher feed rates are obtained with the hopping mode of conveying (that is, when $A_n/g_n > 1.0$). The modes of conveying are summarized in the following flow diagram:

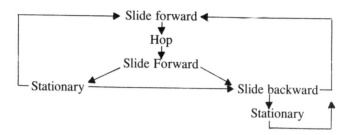

Clearly, a complete analysis of all the possible modes of vibratory conveying is complicated. Such an analysis has been made [1] and leads to equations that must be solved numerically with the aid of a digital computer. For the purposes of the present discussion, it is considered adequate to describe only the main results of this analysis and the results of some experimental tests. In the following, the effects of frequency f, track acceleration A_n/g_n, track angle θ, vibration angle ψ, and the effective coefficient of friction μ on the mean conveying velocity v_m are discussed separately.

3.2 EFFECT OF FREQUENCY

One principal result of the theoretical work is that, for given conditions and for constant track acceleration (that is, A_n/g_n constant), the mean conveying velocity v_m is inversely proportional to the vibration frequency f. Hence,

$$fv_m = \text{const} \tag{3.11}$$

This is illustrated in Fig. 3.4, where the effect of track acceleration on the mean conveying velocity is plotted for three values of the vibration angle ψ. It can be seen that the experimental points for a range of frequencies fall on one line when the factor fv_m is used as a measure of the conveying velocity. This verifies the prediction of the theoretical analysis. One consequence of this result is that, for high conveying velocities and hence high feed rates, it is desirable to use as low a frequency as practicable. However, since the track accelerations must be kept constant, this result means a corresponding increase in track amplitude. The mechanical problems of connecting the feeder to a stationary machine imposes a lower limit on the frequency, but some advantages

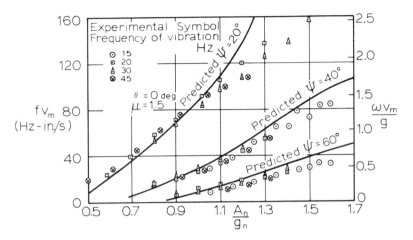

Fig. 3.4 Effect of vibration angle, track acceleration, and frequency on conveying velocity, where ψ is the vibration angle (deg), f the frequency (Hz), θ the track angle (deg), μ the coefficient of friction, and v_m the mean conveying velocity (from Ref. 1).

can be gained by lowering the operating frequency of a bowl feeder from the usual 60 Hz to 30 Hz [1].

3.3 EFFECT OF TRACK ACCELERATION

Figure 3.4 shows that an increase in track acceleration A_n/g_n generally produces an increase in conveying velocity. At some point, however, although the theoretical analysis predicts further increases in velocity, increases in A_n/g_n cease to have a significant effect. This finding may be explained as follows.

If the track acceleration is increased until $A_n/g_n > 1.0$, the part starts to hop once during each cycle, as described earlier. At first, the velocity of impact as the part lands on the track is small but, as the track acceleration is increased further, the impact velocity also increases until, at some critical value, the part starts to bounce. Under these circumstances, the feeding cycle becomes erratic and unstable, and the theoretical predictions are no longer valid.

To obtain the most efficient feeding conditions, it is necessary to operate with values of A_n/g_n greater than unity but below the values that will produce unstable conditions. From Fig. 3.4, it can be seen that, within this range, an approximately linear relationship exists between the factors fv_m and A_n/g_n for each value of ψ and for given values of track angle θ and coefficient of friction μ.

3.4 EFFECT OF VIBRATION ANGLE

From Fig. 3.4, it can be seen that the conveying velocity is sensitive to changes in the vibration angle ψ. The effect is shown more clearly in Fig. 3.5, which indicates that an optimum vibration angle exists for given conditions. For clarity, these theoretical predictions are shown without supporting experimental evidence. Previous work has resulted in the relationship between optimum vibration angle ψ_{opt} and the coefficient of friction, as shown in Fig. 3.6, for a practical value of track acceleration, where A_n/g_n is 1.2.

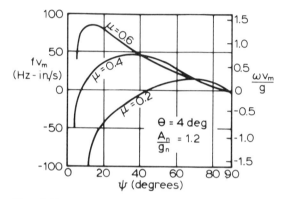

Fig. 3.5 Theoretical results showing the effect of vibration angle on the mean conveying velocity (from Ref. 1).

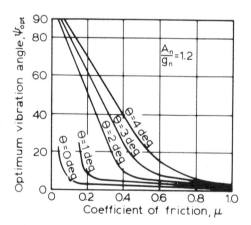

Fig. 3.6 Theoretical results showing the effect of coefficient of friction on the optimum vibration angle (from Ref. 1).

3.5 EFFECT OF TRACK ANGLE

Figure 3.7 gives the effect of track angle θ on the conveying velocity for various track accelerations when μ is 0.2. These results show that the highest velocities are always achieved when the track angle is zero and second, that forward conveying is obtained only with small track angles. The mechanical design of a bowl feeder necessitates a positive track angle of 3–4 deg in order to raise the parts to the bowl outlet. However, it can be seen from the figure that even if conveying can be achieved on the track, the mean conveying velocity will be significantly lower than that around the flat bottom of the bowl. This means that, in practice, the parts on the track will invariably be pushed along by those in the bottom of the bowl, where parts will tend to circulate at a greater speed. This leads to certain problems in the design of the orienting devices, which are generally placed around the upper part of the bowl track.

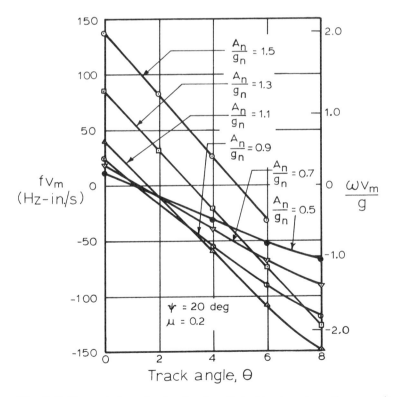

Fig. 3.7 Theoretical results showing the effect of track angle on the conveying velocity (from Ref. 2).

During the testing of such orienting devices, parts transported individually along the track may behave correctly. However, when the bowl is filled and a line of parts forms along the track, the parts tend to be forced through the orienting devices by the pressure of those in the bottom of the bowl. This pressure may often lead to jamming and general unreliability in operation.

From the foregoing discussion, it is clear that, considering the unrestricted feed rate from a bowl feeder, a track angle of 0 deg should be employed because the feeding characteristics in the flat bowl bottom will generally govern the overall performance of the feeder.

3.6 EFFECT OF COEFFICIENT OF FRICTION

The practical range of the coefficient of friction in vibratory feeding is 0.2–0.8. The figure 0.2 represents a steel part conveyed on a steel track. When the track is lined with rubber, a common practice in industry, the coefficient of friction may be raised to approximately 0.8.

Figure 3.8 shows the effect of the coefficient of friction on the conveying velocity for a horizontal track, a vibration angle of 20 deg, and for various track accelerations. It can be seen that, for practical values of track acceleration, an increase in friction leads to an increase in conveying velocity; hence

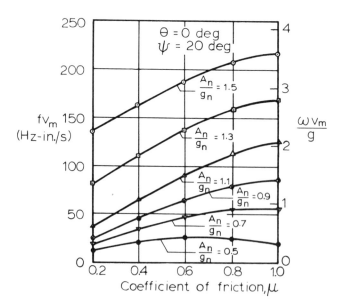

Fig. 3.8 Theoretical results showing the effect of the coefficient of friction on the conveying velocity (from Ref. 2).

the advantage of increasing friction by coating the tracks of bowl feeders with rubber. Coatings can also reduce the noise level resulting from the motion of the parts which is often an important consideration.

3.7 ESTIMATING THE MEAN CONVEYING VELOCITY

At any point on a horizontal track, the ratio of the amplitudes of the vertical and horizontal components of vibration is equal to the tangent of the vibration angle ψ. When the bowl is operating properly, with no rocking motion, the vertical component of motion a_n will be the same at every location in the bowl. The magnitude of the horizontal component a_p, however, changes with radial position.

The horizontal component increases linearly with increasing radial position. If the vibration angle ψ_1 at radial position r_1 is known, the vibration angle ψ_2 at a radial position r_2 is found from

$$\tan \psi_2 = \frac{r_1}{r_2} \tan \psi_1 \qquad (3.12)$$

If the leaf springs are inclined at 60 deg (1.05 rad) from the horizontal plane and attached to the bowl 100 mm from the bowl center, the vibration angle at this radius is the complement of the spring inclination angle, or 30 deg (0.52 rad) if the vibration of the base is neglected. If the vibration of the base is important, the vibration angle should be determined experimentally by comparing the signals from two accelerometers, one mounted vertically and the other horizontally.

For the present example, the magnitude of the vibration angle at a radial position 150 mm from the bowl center can be found from Eq. (3.12) and

$$\psi_2 = \arctan \left[\frac{100}{150} \tan 30\,\text{deg} \right] = 21\,\text{deg}\,(0.37\,\text{rad}) \qquad (3.13)$$

The vibratory motion of a bowl feeder causes parts randomly deposited in the bottom of the bowl to climb the helical track on the interior of the bowl wall. The conveying velocity of the parts on the inclined track is usually governed by the pushing action of the parts circulating around the bottom of the bowl. For those parts moving on the horizontal bottom of the bowl, the conveying velocity v_m depends mainly on the vibration angle ψ, the amplitude of vibration a_0, and the frequency of vibration f or the angular frequency of vibration ω, where ω equals $2\pi f$. A simple dimensional analysis of this situation shows that

$$\frac{v_m \omega}{g} = \text{function} \frac{a_0 \omega^2 \sin \psi}{g}, \psi \qquad (3.14)$$

where g is the acceleration due to gravity and is equal to 9.81 m/s². The functional relationship from Eq. (3.14) is presented graphically in Fig. 3.9. For the usual case of 60-Hz vibration, the conveying velocity is also shown as a function of the vibration angle ψ and the vertical amplitude of vibration, that is, the amplitude normal to the horizontal track. The dimensionless scales shown at the top and at the right of Fig. 3.9 can be used for any vibration frequency including 60 Hz, but this requires additional computation.

For example, suppose that, in a 60-Hz vibratory-bowl feeder, the start of the track, on the bottom of the bowl, is located 150 mm from the bowl center and the vibration angle at this point is 21 deg (0.37 rad). Then, according to Fig. 3.9, the conveying velocity v_m is approximately 36 mm/s when a_n is 80 μm. As a comparison, if a_n or $a_0 \sin \psi$ equals 80 μm, the magnitude of the dimensionless amplitude

$$\frac{a_0 \omega^2 \sin \psi}{g} = \frac{(80 \times 10^{-6})(2\pi \times 60)^2}{9.81} = 1.16 \qquad (3.15)$$

From Fig. 3.9, if ψ equals 21 deg (0.37 rad), the dimensionless velocity $v_m \omega/g$ equals 1.35 and v_m equals 35 mm/s. The corresponding value of the horizontal amplitude a_p is 210 μm.

The vibration amplitude can be adjusted while an operator monitors a special decal mounted on the outer rim of the bowl. This decal is used to measure the peak-to-peak amplitude or twice the horizontal amplitude of vibration at

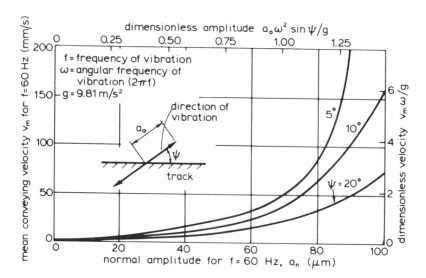

Fig. 3.9 Estimation of the mean conveying velocity on a horizontal track (from Ref. 3).

that point. The correct value of this horizontal amplitude depends on the bowl diameter and is found from geometry.

Using the previous example, if the diameter of the bowl is 600 mm, the parallel amplitude at the rim is

$$\frac{300}{150}(210) = 421\mu m \qquad (3.16)$$

so that the peak-to-peak setting is 0.84 mm. As a consequence, the vibration angle for the last horizontal section of the track is

$$\arctan\left(\frac{80}{420}\right) = 11 \deg (0.19 \,\text{rad}) \qquad (3.17)$$

As the parts leave the bowl, their conveying velocity is now 70 mm/s, from Fig. 3.9.

Although parts are apparently conveyed by vibratory motion with an almost constant conveying velocity, this motion is, actually, a combination of a variety of dissimilar smaller motions giving the total effect of smooth translation. This combination of smaller motions is cyclic and usually repeats with the frequency of the drive. Some of the details of this motion are important in the design of orienting devices used in vibratory-bowl feeders. Figures 3.10 and 3.11 show the effective length J and height H of a hop for a point mass traveling on a horizontal track as shown in Fig. 3.12. The effective length of

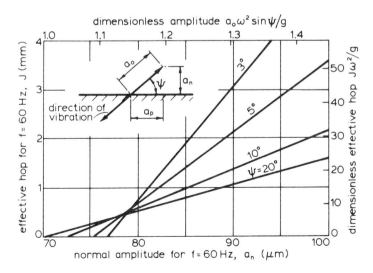

Fig. 3.10 Theoretical estimate of the length of the effective hop on a horizontal track (from Ref. 3).

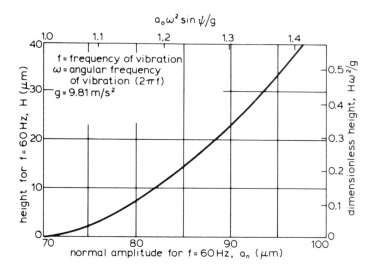

Fig. 3.11 Theoretical estimate of the maximum height reached by a hopping part above a horizontal track (from Ref. 3).

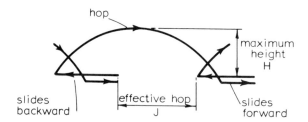

Fig. 3.12 Typical part motion, including the hop.

the hop is the smallest gap in the track that will reject all point masses traveling with this motion.

The magnitude of this effective hop can be determined from Fig. 3.10. If the normal amplitude a_n is 80 μm for 60-Hz vibration and the vibration angle ψ is 11 deg (0.19 rad), then, from this figure, the value of the effective hop J is 0.6 mm. Scales on the top and right side of Fig. 3.10 can be used for other frequencies of vibration, as explained in the discussion of Fig. 3.9.

Similarly, Fig. 3.11 can be used to determine the magnitude of the maximum height H of the hop above the horizontal track. Using the previous conditions ($a_n = 80\mu$m and 60 Hz), H is 8 μm.

It should be noted that intensive theoretical and experimental work has been carried out by Jimbo et al [4]. on the mechanics of vibratory feeding and that they have presented numerous graphs that can be used to estimate the conveying velocity of a part under a wide range of conditions.

3.8 LOAD SENSITIVITY

One of the main disadvantages of vibratory-bowl feeders is their change in performance as the bowl gradually empties. This change occurs because, for a constant power input, the amplitude of vibration and, hence, the maximum bowl acceleration usually increases as the effective mass of the loaded bowl reduces. It can be deduced from Fig. 3.13 that this increase in bowl acceleration will generally result in an increase in the unrestricted feed rate. Vibratory-bowl feeders are often used to convey and orient parts for automatic

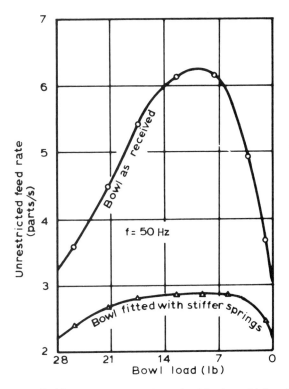

Fig. 3.13 Experimentally determined load sensitivity of a commercial bowl feeder (from Ref. 2).

assembly and, since the workheads on an assembly machine are designed to work at a fixed cycle time, the parts can only leave the feeder at a uniform rate. The feeder must therefore be adjusted to overfeed slightly under all conditions of loading, and excess parts are continuously returned from the track to the bottom of the bowl.

The change in performance as a feeder gradually empties is referred to as its *load sensitivity*, and the upper curve in Fig. 3.13 shows how the unrestricted feed rate for a commercial bowl feeder in the as-received condition varied as the bowl emptied. It can be seen that the maximum feed rate occurred when the bowl was approximately 25% full and that this represented an increase of approximately 100% of the feed rate obtained with the bowl full. It is of interest to compare this result with the measured changes in bowl acceleration show in Fig. 3.14, where it can be seen that the bowl acceleration and, hence, the amplitude increased continuously until the bowl became empty. Clearly, when a feeder empties, the feed rate will reduce to zero, but Fig. 3.13 shows that the feed rate begins to reduce much sooner than might be expected from Fig. 3.14. This behavior is considered to be due to the greater velocity of parts in the flat bowl bottom than that on the track; this was described earlier. When the bowl is full, the feed rate depends mainly on the feeding characteristics in the bottom of the bowl, where the general circulation of parts pushes those on the track. However, as the bowl empties, leaving mainly those parts that are held on the track, the pushing action ceases, and the feed rate depends on the conveying velocity on the inclined track, which is generally lower than that on a horizontal surface. This explains the difference in character between the graphs in Figs. 3.13 and 3.14.

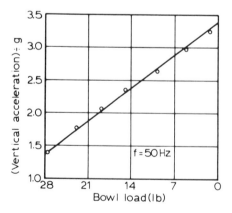

Fig. 3.14 Effect of bowl load on bowl acceleration (from Ref. 2).

Figure 3.13 suggests that, under the test conditions, the as-received bowl feeder could be used to feed a workhead operating at a maximum rate of 3 insertions per second and that, in this case, there would be considerable recirculation of parts resulting from overfeeding. Assuming that the feeder is to be refilled when it becomes 25% full, the feeding characteristics between refills may be reasonably represented by a feed rate increasing linearly as the bowl empties.

3.9 SOLUTIONS TO LOAD SENSITIVITY

One of the simplest solutions to load sensitivity and the one most commonly used is the load detector switch together with a secondary feeder. The load detector switch is simply a mechanical arm and limit switch that detects when the level of parts in the bottom of the bowl falls below some predetermined level. When closed, the switch activates the secondary feeder and refills the bowl to the predetermined level. This action essentially increases the frequency of refills, reducing the recirculation effect to almost zero.

A second solution requires modification to the feeder. Frequency-response curves for the vibratory bowl feeder used in the previous experiments are presented in Fig. 3.15. These curves show the effect of changes in the forcing frequency on the bowl acceleration for a constant power input and for various bowl loadings. In these tests, the power input is less than that employed for the results in Fig. 3.14, but they show the same effect where, for a forcing frequency of 50 Hz, the maximum bowl acceleration is sensitive to changes in bowl loading. However, it can also be seen that, whereas for a forcing frequency of 50 Hz (the frequency used in Great Britain), the maximum bowl acceleration is sensitive to changes in bowl loading, for a forcing frequency of approximately 44 Hz, the bowl acceleration is approximately constant for all bowl loadings. Under these latter conditions, the load sensitivity would be considerably reduced. Alternatively, it is clear that increasing the spring stiffness of the bowl supports sufficiently would have the effect of shifting the response curves to the right and minimizing the changes in bowl acceleration for a forcing frequency of 50 Hz. The natural frequency of the empty, as-received bowl was approximately 53 Hz, and tests showed that, if this was increased to 61 Hz by increasing the support spring stiffness, the load sensitivity of the feeder was considerably reduced. The lower curve in Fig. 3.13 shows this effect. It can also be seen, however, that the feed rates have been reduced by stiffening the support springs and, therefore, in order to maintain the higher feed rate, a more powerful drive would be required. Generally, vibratory-bowl feeders are tuned to a natural frequency just slightly higher than the frequency of the drive to minimize the power by utilizing the ease of transmitting vibration at or near the natural frequency.

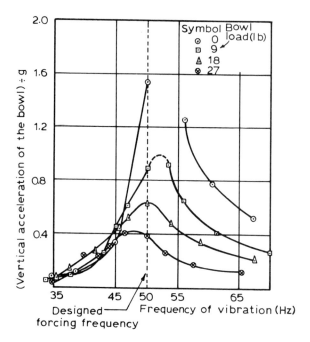

Fig. 3.15 Frequency-response curves for a vibratory-bowl feeder, showing the effect of bowl load (from Ref. 2).

A third solution uses on/off controls in the feed track or delivery chute to control the operation of the feeder. A line of parts is stored in the external feed track and, when the line becomes small, the lower sensor activates the feeder, filling the line to the upper sensor which, in turn, shuts the feeder off.

Some of the more expensive vibratory-bowl feeders use silicon-controlled rectifier (SCR) drive systems, which can be coupled with accelerometer feedback to maintain a constant amplitude of vibration. This prevents the mean conveying velocity of the parts from increasing as the bowl empties.

3.10 SPIRAL ELEVATORS

A device commonly employed for elevating and feeding component parts is the spiral elevator. A typical spiral elevator is illustrated in Fig. 3.16; it can be seen that the drive is identical to that used for a vibratory bowl feeder. The helical track passes around the outside of a cylindrical tube. This device is not generally used to orient parts because the parts cannot readily be rejected back into the hopper bowl situated at the base of the elevator. Since the mode of conveying for the parts is identical to that obtained with a vibratory-bowl

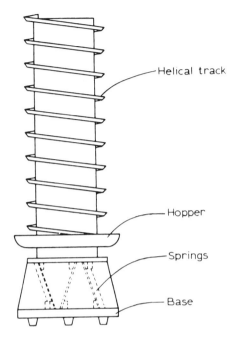

Fig. 3.16 Spiral elevator.

feeder, the results and discussion presented earlier and the design recommendations made will also apply to the spiral elevator.

3.11 BALANCED FEEDERS

One of the major problems in the application of vibratory feeding is the need to isolate the feeder vibrations from the structure of the assembly machine. For this reason, the base of a vibratory-bowl feeder is usually supplied with rubber feet. However, these feet provide the feeder with additional degrees of freedom and, if the bowl and tooling are not dynamically balanced, a rocking motion can be superimposed on the motions of the bowl. This rocking motion affects the vibration angle at the bowl track so that, in some regions of the track, the vibration angle is increased and, in some regions, it is decreased. This results in erratic behavior of the parts so that, along certain portions of the track, a high conveying velocity occurs while, along others, the parts slow down or even tend to stop.

The normal solution to this problem is to ensure that the bowl and its tooling are properly balanced. However, Yokoyama et al [5]. have proposed a variety of vibratory feeder designs in which the horizontal components of the inertia forces cancel, resulting in a balanced design that minimizes the vibrations transmitted through the feeder supports. This approach is most easily described by referring to a recirculating vibratory feeder consisting of two linear vibratory tracks mounted side by side. Figure 3.17a shows a simple horizontal linear vibratory track, and it can be seen that the vertical components of any inertia forces would be carried equally by the two feet, whereas horizontal components would superimpose a rocking motion. This rocking motion has the effect of increasing the conveying velocity at the beginning of the track and reducing it at the end of the track.

Figure 3.17b shows two such tracks mounted side by side in such a way that the horizontal components of the inertia forces cancel. Not only does this eliminate the rocking motion, but it provides a means of recirculating parts so that we have an alternative to the vibratory-bowl feeder, and one that enables

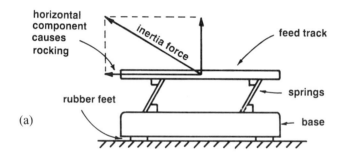

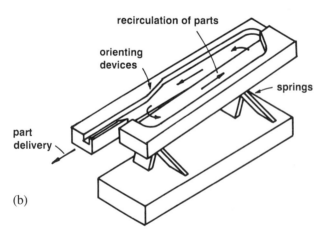

Fig. 3.17 Balanced vibratory feeder.

us to place orienting devices on one of the tracks. Thus, parts in the correct orientation will pass through the devices to be delivered to the end of the track, whereas other parts are rejected onto the return track.

Commercial versions of this balanced feeder arrangement are available that are suitable for very small parts, such as surface-mount resistors. This balancing principle has also been applied successfully to a variety of vibratory-bowl feeder arrangments [4].

3.12 ORIENTATION OF PARTS

In an automatic assembly machine, the parts must be fed to the workheads correctly oriented. The devices employed to ensure that only correctly oriented parts are fed to the workhead fall into two groups: those that are incorporated in the parts feeder, which are usually referred to as *in-bowl tooling*; and those that are fitted to the chute between the feeder and the workhead, called *out-of-bowl tooling*. The devices used for in-bowl tooling very often work on the principle of orienting by rejection and may be termed *passive orienting devices*. With this type of device, only those parts that, by chance, are fed correctly oriented pass through the device, while the other parts fall back into the hopper or bowl. The rejected parts are then refed and make a further attempt to pass through the orienting devices. In some cases, devices are fitted that reorient parts. These may be termed *active orienting devices* and, although they are not as widely applicable, they have the advantage that no reduction in feed rate occurs as a result of the rejection of parts that have already been fed. Some orienting devices are fitted between the parts feeder and the automatic workhead. Since, with this system, rejected parts cannot easily be returned to the parts feeder, orienting devices employed in this way are usually of the active type.

The following describes some of the more common orienting devices and tooling employed in feeders for automatic assembly.

3.13 TYPICAL ORIENTING SYSTEM

Of all the various types of feeding devices, vibratory-bowl feeders allow by far the greatest flexibility in the design of orienting devices. Figure 3.18 shows the orienting system commonly employed to orient screws in a vibratory-bowl feeder. In this arrangement, the first device, a wiper blade, rejects all the screws not lying flat on the track. The gap below the blade is adjusted so that a screw standing on its head or a screw resting on the top of others is either deflected back into the bowl or deflected so that the screw lies flat on the track. Clearly, the wiper blade can be applied here only if the length of the screw is greater than the diameter of its head. The next device, a pressure break, allows

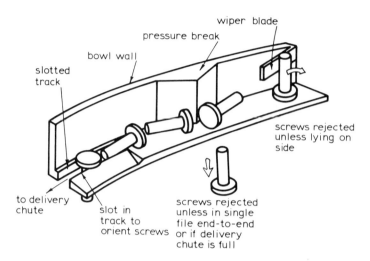

Fig. 3.18 Orientation of screws in vibratory-bowl feeder.

screws to pass only in single file and only with either head or shank leading. Screws being fed in any other attitude will fall off the narrow track and back into the bowl at this point. The pressure break also performs another function; if the delivery chute becomes full, excess parts are returned to the bottom of the bowl at the pressure break, and congestion in the chute is therefore avoided. The last device consists of a slot in the track that is sufficiently wide to allow the shank of the screw to fall through while retaining the screw head. Screws arriving at the slot either with the shank leading or with the head leading are therefore delivered with the shank down, supported by the head. In this system for orienting screws, the first two devices are passive and the last is active.

Although the devices described above are designed for a certain shape of part, two of them have wide application in vibratory-bowl feeding. First, a pressure break is usually necessary because most feeders are adjusted to overfeed slightly in order to ensure that the workhead is never "starved" of parts. With this situation, unless a level-sensing device controlling the feeder output is attached to the delivery chute, the delivery chute is always full, and a pressure break provides a means of preventing congestion at its entrance.

Second, the wiper blade is a convenient method of rejecting parts that are resting on top of others. In a vibratory-bowl feeder, this rejection often occurs because of the pushing action of parts traveling up the track. However, care must be taken in applying the wiper blade because thin parts may have a tendency to jam under the blade, as illustrated in Fig. 3.19. The tendency for this

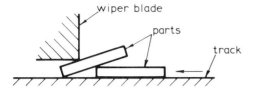

Fig. 3.19 Thin parts jammed under wiper blade.

jamming to occur will be reduced by arranging the blade so that it lies at an acute angle to the bowl, as shown in Fig. 3.18. In some cases, an alternative approach is necessary, and this is illustrated in Fig. 3.20, which shows the orienting device commonly employed to orient washers. It can be seen that a portion of the track is arranged to slope sideways and down toward the center of the bowl. A small ledge is provided along the edge of this section of the track to retain those washers that are lying flat and in single file. Other washers will slide off the track and into the bowl. With a device of this type, where the parts are rotated as they are fed, it is often necessary to arrange the design of the track to ensure that the path of the center of gravity of the part is not raised rapidly; otherwise, a serious reduction in feed rate may occur.

Figure 3.21 shows a refinement made in the orienting device described above. In this case, machined washers may be oriented by providing a ledge sufficiently large to retain a washer being fed base down (Fig. 3.21a) but too small to retain a washer being fed base up (Fig. 3.21b).

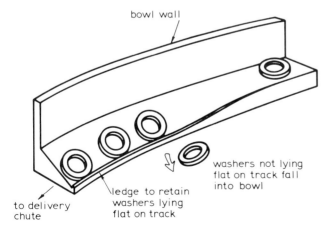

Fig. 3.20 Orientation of washers in vibratory-bowl feeder.

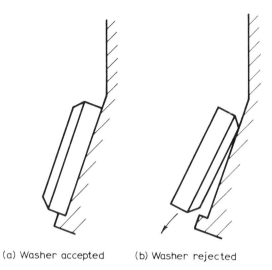

(a) Washer accepted   (b) Washer rejected

Fig. 3.21 Orientation of machined washers.

Figure 3.22 illustrates a common type of orienting device known as a *cutout*, where a portion of the track has been cut away. This device makes use of the difference in shape between the top and the base of the part to be fed. Because of the width of the track and the wiper blade, the cup-shaped part can only arrive at the cutout resting on its base or on its top. It can be seen from

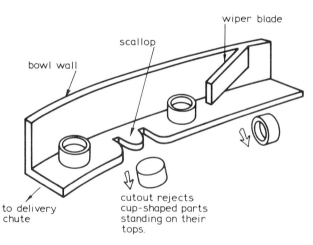

Fig. 3.22 Orientation of cup-shaped parts in vibratory-bowl feeder.

Fig. 3.22 that the cutout has been designed so that a part resting on its top falls off the track and into the bowl, whereas a part resting on its base passes over the cutout and moves on to the delivery chute.

Figure 3.23 shows another application of a cutout, in which the area covered by the top of a part is very much smaller than the area covered by its base. In this case, a V-shaped cutout rejects any part resting on its top.

Figure 3.24 shows an example in which U-shaped parts are oriented. With parts of this type, it is convenient to feed them supported on a rail. In this

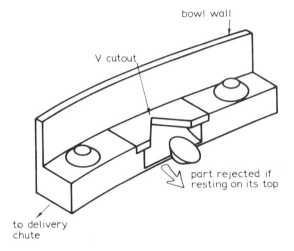

Fig. 3.23 Orientation of truncated cones in vibratory-bowl feeder.

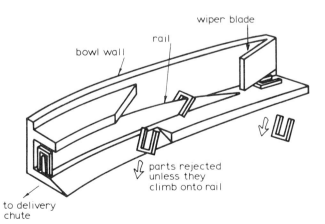

Fig. 3.24 Orientation of U-shaped parts in vibratory-bowl feeder.

case, a portion of the parts climb onto the rail and pass to the delivery chute. The remainder fall either directly into the bowl or into the bowl through a slot between the rail and the bowl wall.

Figure 3.25 shows a narrowed-track orienting device that is generally employed to orient parts lengthwise end to end while permitting only one row to pass. Finally, Fig. 3.26 shows a wall projection and narrowed-track device used to feed and orient parts with steps or grooves, such as short, headed

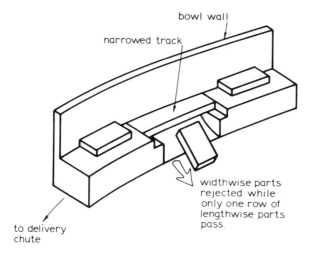

Fig. 3.25 Narrowed track.

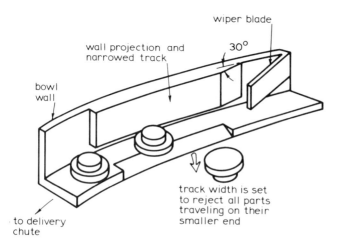

Fig. 3.26 Wall projection and narrowed track.

parts. A short, headed part traveling on its larger end passes through the device, but other orientations are rejected back into the bowl.

3.14 EFFECT OF ACTIVE ORIENTING DEVICES ON FEED RATE

Sometimes, a part used on an assembly machine has only a single orientation but, more often, the number of possible orientations is considerably greater. If, for example, a part had eight possible orientations and the probabilities of the various orientations were equal and, further, if only passive orienting devices were used to orient the parts, the feed rate of oriented parts would be only one-eighth of the feed rate of unoriented parts. It is clear that if active orienting devices could be utilized, the feed rate of oriented parts could be considerably increased.

To illustrate this point, consider the orienting system shown in Fig. 3.27 for feeding rectangular blocks. At one point, the width of the track is such that blocks can be fed only with their long axes parallel to the direction of motion. Also, a wiper blade is arranged so that blocks lying flat or standing on their sides will be accepted. It is assumed in this example that the width of a block is less than twice its thickness. Finally, an active orienting device in the form of a tapered element ensures that blocks lying flat will be turned to stand on their sides. With this arrangement, all the blocks fed up the track with their long axes parallel to the conveying direction will be fed from the bowl. If, however, the active orienting device was not part of the system and the wiper

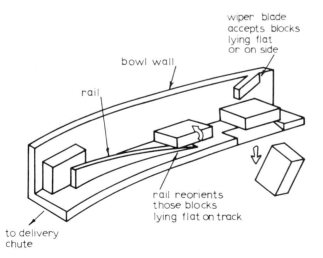

Fig. 3.27 Orientation of rectangular blocks in vibratory-bowl feeder.

blade had been arranged to accept only the blocks lying flat on the track, a larger proportion of blocks would have been rejected, with a consequent reduction in feed rate.

3.15 ANALYSIS OF ORIENTING SYSTEMS

To determine the effect of certain aspects of part geometry on the efficiency with which it can be fed and oriented, consider several parts having the same basic shape but different sizes [6]. Figure 3.28 shows a family of parts that take the form of plain cylinders, with a blind hole drilled axially from one end. These parts have the same diameter, 12.7 mm, with an 11.7-mm-diam. square bottom hole drilled from one end to a depth of 0.718 times the length of the parts. It can be shown that, for all these parts, the center of mass is positioned at the bottom of the hole; the only geometric variable necessary to describe the part is its length-to-diameter ratio.

For a bowl feeder having a track that ensures single-file feeding of these parts, only four orientations of the part need to be considered. These orientations are shown in Fig. 3.29 and are keyed a, b_1, b_2, and c. Orientation a shows the part fed standing on its base (that is, heavy end down). Orientations b_1 and b_2 show the part fed on its side, either heavy end first (b_1) or light end first (b_2). Finally, orientation c shows the part fed heavy end up.

Before a study of the design of an orientation system for these parts can be made, it is necessary to know the probabilities with which these four orientations would initially occur. This information can then be used as the input to the orienting system analysis. Figure 3.30 presents the results of experimental and theoretical work. In the experiments, each part was repeatedly thrown onto a flat horizontal aluminum surface, and the resulting final resting aspect was noted for each trial. In this experiment, it is not possible to distinguish

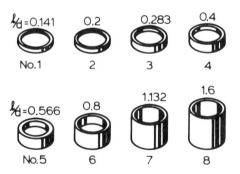

Fig. 3.28 Parts used in the experiment (from Ref. 6).

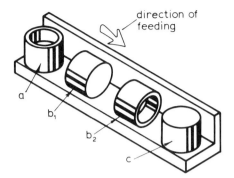

Fig. 3.29 Orientations of cup-shaped parts (from Ref. 6).

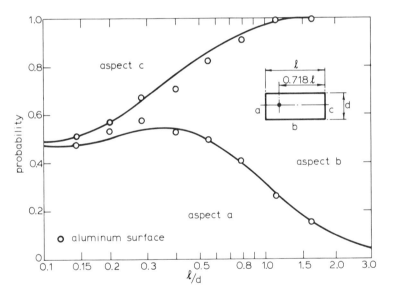

Fig. 3.30 Distribution of natural resting aspects (from Ref. 6).

between orientations b_1 and b_2 because the direction of feeding is not defined. For this reason, the term *natural resting aspect* is employed, which is meant to describe the way in which a part can rest on a horizontal surface. Thus, natural resting aspect a designates parts resting on their bases, b parts resting on their sides, and c parts resting with their heavy ends up. Those parts that come to rest on their sides will, when fed, divide about equally into the two orientations b_1 and b_2 described above. It can be seen from Fig. 3.30 that, for example, the probability that a cup-shaped part having a length-to-diameter ratio of 0.8 will come to rest on its base (natural resting aspect a) is 0.4. Each experimental point in Fig. 3.30 represents the results of at least 250 trials and is subject to 95% confidence limits of less than ± 0.05.

The basis of a theoretical study of the distributions of natural resting aspects of parts is described later in this chapter. The solid lines shown in Fig. 3.30 are theoretical curves and can be seen to agree closely with the experimental results.

3.15.1 Orienting System

A system for orienting the cup-shaped parts is shown in Fig. 3.31. It consists of one active device, a step; and two passive devices, a scallop and a sloped track with a ledge. The system is designed to deliver a part in orientation a (on its base).

In the orientation process, the parts first encounter the step device, whose purpose is to increase the proportion of parts in orientation a. This increase is achieved by arranging a step height that does not affect many of the parts in orientation a but reorients some of the parts in orientations b_1 or c into orientation a as they pass over the step.

The remaining passive devices are simply designed to ensure that all parts remaining in orientations b_1, b_2, and c are rejected back into the bowl. These rejected parts later make a further attempt to filter through the orienting system.

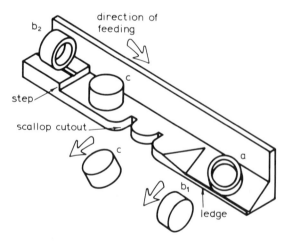

Fig. 3.31 Experimental orienting system (from Ref. 6).

The first of the passive devices, the scallop cutout, ensures rejection of a part in orientation c. Its design was based on data regarding the feeding motion of the part when the horizontal amplitude of vibration at the bowl wall was set at 1.0 mm with a vibration angle of 5 deg. The part motion under these conditions is shown in Fig. 3.32 and was obtained using a computer program similar to that described by Redford and Boothroyd [1]. It can be seen from the figure that the part slides forward a distance of 0.86 mm after hopping and then slides backward a distance of 0.91 mm before hopping again. Although the distance the part hops is 2.72 mm, the maximum gap in the track that any point on the undersurface of the part can negotiate is 1.75 mm. With this information, it is possible to design the scallop device so that parts in orientation a are always supported, regardless of where they are situated on the device, whereas those parts in orientation c will, at some point, be situated in a position where they cannot be supported and will fall off the track into the bowl. Experiments [6] show that a small proportion of those parts in orientations b_1 and b_2 are also rejected by this device, but this does not affect the performance of the system since the next device will reject parts in both these orientations.

The second passive device, the sloped track with a ledge, is designed to reject all those parts in orientations b_1 and b_2. The ledge retains parts in orientation a but does not prevent parts in orientations b_1 and b_2 from rolling off the track and back into the bowl.

The only orienting system variable considered here is the height of the active step orienting device and, in order to perform an optimal design analysis, it is necessary to carry out an experimental program to measure the effect of various step heights on each orientation of each of the eight specimens. Typical results of such experiments are presented in Fig. 3.33, which shows the effects of feeding a part in each of its four initial orientations, over the step, with step heights varying from zero to a maximum of 7 mm. It was

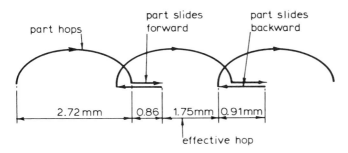

Fig. 3.32 Motion of part relative to track during experiments (from Ref. 6).

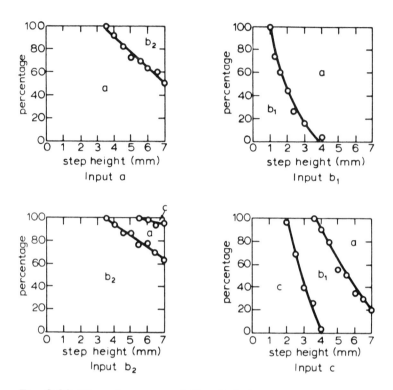

Fig. 3.33 Effect of step on part 7 (from Ref. 6).

found that, for step heights greater than 7 mm, the parts would bounce errati-
cally upon landing on the track below the step, an effect that would be unacceptable in practice.

3.15.2 Method of System Analysis

The object of an analysis of a bowl-feeder orienting system is to design each
device so that the highest value for the efficiency of the complete system is
obtained. The efficiency of the system is defined as the number of properly
oriented parts delivered by the system divided by the number of parts entering
the system.

To calculate this efficiency for a system of orienting devices, a matrix technique has been developed [7]. Each device is represented by a matrix whose
number of rows and columns depends on the number of orientations in the
devices's respective input and output. If these matrices are multiplied in the
order in which the parts encounter the devices, the resulting single-column

matrix represents the performance of the system. If this matrix is then premultiplied by a single-row matrix representing the initial distribution of orientations of the part, the efficiency of the system is obtained.

Figure 3.34 shows a schematic diagram for the present system, together with the appropriate matrices. The terms in the matrix that represent the step device are only symbolic. The term AA indicates the proportion of those parts in orientation a that will remain in orientation a, $AB1$ represents those parts in orientation a that are reoriented into b_1, and so on. In the matrix for the scallop device, q_1 and q_2 represent the proportion of parts that enter the device in orientations b_1 and b_2, respectively, and exit in the same orientation. For part 7 and a step height of 7 mm, the step orienting device matrix becomes

$$
\begin{array}{c}
a \\
b_1 \\
b_2 \\
c
\end{array}
\begin{bmatrix}
\begin{array}{cccc}
a & b_1 & b_2 & c \\
0.50 & 0 & 0.50 & 0 \\
1.00 & 0 & 0 & 0 \\
0.30 & 0 & 0.64 & 0.06 \\
0.80 & 0.20 & 0 & 0
\end{array}
\end{bmatrix}
\qquad (3.18)
$$

and the resulting system matrix is

$$
\begin{array}{c}
a \\
b_1 \\
b_2 \\
c
\end{array}
\begin{bmatrix}
\begin{array}{cccc}
a & b_1 & b_2 & c \\
0.50 & 0 & 0.50 & 0 \\
1.00 & 0 & 0 & 0 \\
0.30 & 0 & 0.64 & 0.60 \\
0.80 & 0.20 & 0 & 0
\end{array}
\end{bmatrix}
\begin{bmatrix}
\begin{array}{ccc}
a & b_1 & b_2 \\
1 & 0 & 0 \\
0 & q_1 & 0 \\
0 & 0 & q_2 \\
0 & 0 & 0
\end{array}
\end{bmatrix}
\begin{bmatrix}
a \\
1 \\
0 \\
0 \\
0
\end{bmatrix}
=
\begin{bmatrix}
a \\
0.50 \\
1.00 \\
0.30 \\
0.80
\end{bmatrix}
\qquad (3.19)
$$

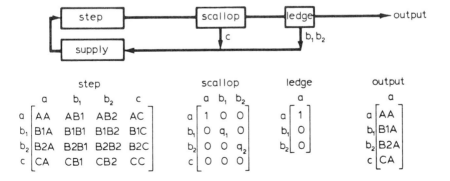

Fig. 3.34 Orienting system analysis (from Ref. 6)

This result means that 50% of those parts that enter the system in orientation a exit in orientation a, 100% of those parts that enter in orientation b_1 exit in orientation a, and so on.

From Fig. 3.30, the initial distribution matrix for part 7 is

$$[a \ b_1 \ b_2 \ c] = [0.27 \ 0.35 \ 0.35 \ 0.03] \qquad (3.20)$$

Thus, premultiplying the system matrix, Eq. (3.19), by the input distribution matrix, Eq. (3.20), gives

$$[0.27 \ 0.35 \ 0.35 \ 0.03] \begin{bmatrix} 0.50 \\ 1.00 \\ 0.30 \\ 0.80 \end{bmatrix} = 0.61$$

which means that, under these conditions, the efficiency of the system is 61%. Hence, if the bowl was set to feed parts at a rate of 10 parts/min, the mean delivery rate of parts in orientation a would be 6.1 parts/min.

3.15.3 Optimization

To optimize the design of this system, it is necessary to determine the step height that gives the maximum efficiency for each of the eight parts. The simplest method is to calculate the system efficiently for increments of step height within the practical range. The results of this procedure are shown in Figs. 3.35 and 3.36, where it can be seen that the maximum efficiency for five of the parts occurs at the maximum allowable step height of 7 mm. These maxima are plotted in Fig. 3.37 (curve B) and compared with the initial distribution of parts in orientation a (curve A). Since this initial distribution is the same as the system efficiency that would be obtained if the step device had not been included in the system, the figures show clearly the advantages of including the step device. In many cases, the efficiency of the system is almost doubled; in other words, the output rate of oriented parts would be almost doubled. Since such significant advantages of including the step device in the system can be demonstrated, it is of interest to consider the effect of including a further step device. In this case, a further variable is introduced, and the upper curve shown in Fig. 3.37 is obtained. In all cases, it is seen that the overall maximum efficiency occurs for a part having a length-to-diameter ratio ℓ/d of 0.4.

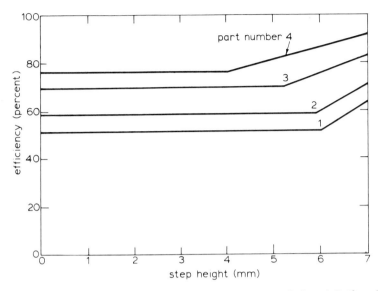

Fig. 3.35 Effect of step height on efficiency (parts 1, 2, 3, and 4) (from Ref. 6).

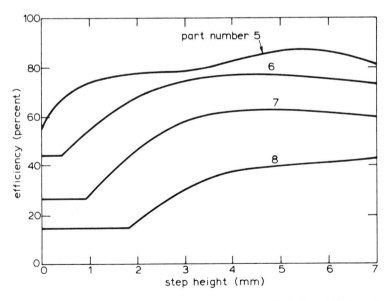

Fig. 3.36 Effect of step height on efficiency (parts 5, 6, 7, and 8) (from Ref. 6).

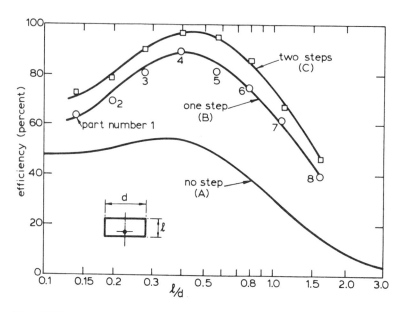

Fig. 3.37 Effect of part shape on efficiency of orienting system (from Ref. 6).

3.16 PERFORMANCE OF AN ORIENTING DEVICE

In the preceding section, to optimize the design of the orienting system for cup-shaped parts, it was necessary to determine the step height giving the maximum efficiency. The information employed to determine the performance of the step orienting device was empirical. It would be most useful in investigations involving the optimization of an orienting system if theoretical expressions were available that described with sufficient accuracy the performance of the orienting device. The present section describes the analysis of one of the passive orienting devices introduced earlier: a V cutout (Fig. 3.23) applied to the orientation of truncated cone-shaped parts. It can be seen from Fig. 3.23 that, with a properly designed device, those parts being fed base uppermost will be rejected, whereas those parts being fed base down will be accepted and allowed to proceed to the outlet chute.

Since the height of the part has no effect on the performance of the device, the only parameters necessary to describe its important characteristics are the radius of the base R and the radius of the top r. The symmetrical orienting device may also be described completely by only two parameters: the half-angle of the cutout θ and the distance b from the apex of the cutout to the bowl wall.

During vibratory feeding, the part proceeds along the track by a combination of discrete sliding motions either backward or forward or both and, under certain conditions, by a forward hop. All the motions occur sequentially during each cycle of the vibratory motion of the bowl. During each cycle, when the conditions are such that the part hops, there will usually be a distance along the track, denoted by J, where the part does not touch the track. Therefore, J is the smallest gap or slot in the track that will reject all particles that travel with this particular motion. For vibratory conditions that produce relatively small sliding motions compared to the hop, the motion can be characterized by a series of equal hops, each a distance J.

The object of the design of a V cutout would be to determine the values of the parameters θ and b such that, for a given part (given values of R and r) and for a given feeding characteristic (given J), all the parts fed on their tops would be rejected, and a maximum of those fed on their bases would be accepted.

3.16.1 Analysis

Figure 3.38 shows two limiting conditions for the position of a part resting on its top. In the first position, the center of the part lies at P on the edge of the cutout. Thus, if the part comes to rest momentarily just to the right of P, it will be rejected. The second limiting condition places the center of the part at Q, and the edge of the part is just supported at D by the edge of the cutout. Thus, if the part contacts the track with its center anywhere between P and Q, it will be rejected. Two similar limiting conditions not shown in the figure will

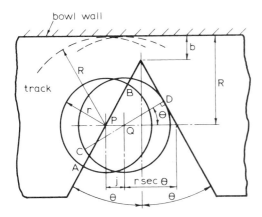

Fig. 3.38 Determination of j for small cutout angles.

exist to the right of the cutout centerline, and these positions may be deduced from the symmetry of the situation.

It is clear that, for a part traveling from left to right (Fig. 3.38) in a series of hops, the probability that its center will fall in the spaces between P and Q (and thus be rejected) is given by j/J, where j is the distance between P and Q, and J is the length of each hop.

For those parts that negotiate the gap between P and Q, the probability that they will clear the first gap is $(1 - j/J)$ and the probability that they will be rejected in the similar gap lying to the right of the cutout centerline is (j/J). Hence, the total probability R_e that a part will be rejected in one of the two gaps is

$$R_e = j/J + (1 - j/J)j/J = 2(j/J) - (j/J)^2 \tag{3.21}$$

For the conditions illustrated in Fig. 3.38,

$$j = 2(R - b)\tan\theta - r\sec\theta \tag{3.22}$$

For large cutout angles, Eq. (3.22) does not always apply because when the top of the part is supported at D by the right-hand edge of the cutout, point C, diametrically opposite D, may be to the right of the left-hand edge of the cutout. Thus, the part in this situation will be rejected, and an alternative limiting condition arises. In this case, point P is unchanged, but point Q is found by arranging for the diameter CD of the part to be just supported between the edges of the cutout. From Fig. 3.39,

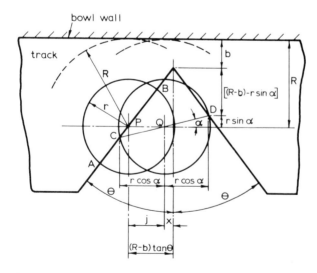

Fig. 3.39 Determination of j for large cutout angles.

$$j = (R - b) \tan \theta - x \tag{3.23}$$

and

$$x = r \cos \alpha - [(R - b) - r \sin \alpha \tan \theta] \tag{3.24}$$

Also,

$$x = [(R - b) + \sin \alpha] \tan \theta - r \cos \alpha \tag{3.25}$$

Eliminating α from Eqs. (3.24) and (3.25) gives

$$x = \tan \theta [r^2 - (R - b)^2 \tan^2 \theta]^{1/2} \tag{3.26}$$

Finally, substituting Eq. (3.26) into Eq. (3.23) gives

$$j = \tan \theta \{(R - b) - [r^2 - (R - b)^2 \tan^2 \theta]^{1/2}\} \tag{3.27}$$

The value of b at which Eq. (3.22) becomes invalid and Eq. (3.27) must be applied can be found by arranging for CD in Fig. 3.39 to lie at right angles to the right-hand edge of the cutout. Under these conditions, α becomes equal to θ, and eliminating x from Eqs. (3.24) and (3.25) gives

$$(R - b) = r \cos \theta \cot \theta \tag{3.28}$$

and, thus, from Eq. (3.22) or (3.27),

$$j = r(2 \cos \theta - \sec \theta) \tag{3.29}$$

It can readily be shown that, for $\theta \geqslant 45$ deg, Eq. (3.27) always applies.

It is convenient to eliminate one variable from the foregoing expressions by dividing through by R and thereby writing the parameters in dimensionless form. Defining $r_0 = r/R$, $b_0 = b/R$, $j_0 = j/R$, and $J_0 = J/R$, the following equation is obtained for the rejection R_e of parts:

$$R_e = 2(j_0/J_0) - (j_0/J_0)^2 \tag{3.30}$$

where

$$j_0 = \tan \theta \{(1 - b_0) - [r_0^2 - (1 - b_0)^2 \tan^2 \theta]^{1/2}\} \tag{3.31}$$

unless $\theta < 45$ deg and $b_0 > (1 - r_0 \cos \theta \cot \theta)$, in which case,

$$j_0 = 2(1 - b_0) \tan \theta - r_0 \sec \theta \tag{3.32}$$

These equations are presented in graphic form in Fig. 3.40 for a cutout having a half-angle of 30 deg and for a part where r_0 is 0.8.

The figure shows how the theoretical rejection rate R_e varies as the parameter b_0 is changed for a given value of the distance hopped J_0 by the part during each vibration cycle. The figures for the case in which the part is being fed on its base were obtained by setting r_0 equal to unity. A negative b refers to the case in which the apex of the cutout lies outside the interior surface of the bowl wall.

It can be seen from the figure that, as the parameter b_0 is gradually decreased, a condition is eventually reached at which the unwanted parts (those being fed on their tops) will start to be rejected. On further decreases in b_0, a point will be reached at which all these parts will be rejected. Similar situations will arise for the wanted parts but for lower values of b_0. Eventually, a value of b_0 will be reached at which all the parts will be rejected. Clearly, in practice, it will be necessary to choose a situation in which all the unwanted parts will be rejected, even at the expense of rejecting some of the wanted parts.

After defining the largest value of b_0 for which all the unwanted parts are rejected as b_u and the smallest value of b_0 at which all wanted parts are accepted as b_w, it can be stated that the best conditions would be those that resulted in the largest value of $(b_u - b_w)$. This would give the greatest working range for a given part and for given feeding conditions.

It is of interest to study how the magnitudes of b_u and b_w are affected by changes in the design parameters.

The value of b_u is obtained by setting R_e equal to unity in Eq. (3.30) with the appropriate equation for j_0, (3.31) or (3.32); the value of b_w is obtained by setting R_e equal to zero and r_0 equal to unity.

Thus, after rearrangement, when $\theta \geqslant 45$ deg,

$$b_u = 1 - \cos^2\theta[J_0 \cot\theta + (r_0^2 \sec^2\theta - J_0^2)^{1/2}] \tag{3.33}$$

$$b_w = 1 - \cos\theta \tag{3.34}$$

when $\theta \leqslant 45$ deg,

$$b_u = 1 - 0.5J_0 \cot\theta - 0.5r_0 \operatorname{cosec}\theta \tag{3.35}$$

$$b_w = 1 - 0.5 \operatorname{cosec}\theta \tag{3.36}$$

unless $b_u < (1 - r_0 \cos\theta \cot\theta)$, in which case it is given by Eq. (3.33) and b_w is given by Eq. (3.36). The equations above are plotted in Fig. 3.41 and illustrate the effects of θ and J_0.

This theory has been developed for an idealized situation in which the part proceeds along the track by hopping. However, in reality, both hopping and sliding occur. It can be shown that, although this would affect Eq. (3.30), and hence the shapes of the curves in Fig. 3.40, the values of b_u and b_w would be unchanged.

Values of the working range $(b_u - b_w)$ obtained from Eqs. (3.33–3.36) are plotted against θ in Fig. 3.42. It can be seen that, for larger values of J_0, an optimum condition exists that gives the maximum working range. Further, at low values of θ, the magnitude of the working range becomes very sensitive to changes in J_0. Since most vibratory feeders operate at the same frequency, a large value of J_0 implies a high conveying velocity. In practice, it would

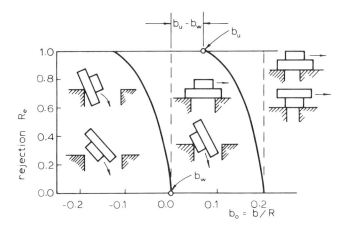

Fig. 3.40 Effect of b_0 on rejection of parts: $\theta = 30$ deg, $J_0 = 0.15$.

clearly be desirable to choose conditions that give minimum sensitivity to changes in the feeding parameters and yet give the maximum working range for a reasonably large value of J_0.

The procedure used to obtain the experimental points for b_u and b_w, which are presented in Fig. 3.41, is outlined in Boothroyd and Murch [8]. It is seen in Fig. 3.41 that the results for b_u show good agreement with the theory over the whole range of cutout angles when J_0 is set equal to 0.15. The experimental values for b_w show good agreement with the theory only at small cutout angles. For larger cutout angles, the experimental value is always the larger.

Ideally, in the design of a V-cutout orienting device, the pertinent data regarding the vibrating motion of the bowl feeder could be used to estimate the value of J_0, employing the results presented earlier in this chapter. Subsequently, using Fig. 3.42, the half-angle of the cutout θ that gives the best value for the working range could be chosen. From this figure, when $J_0 \leqslant 0.1$, the smaller the value of θ, the larger the working range. However, small cutout angles can present a practical problem. Under these circumstances, parts that are rejected may not be deflected properly from the bowl track and may interfere with the behavior of the following parts. Thus, the angle chosen should be that which gives the best working range and yet provides for adequate deflection of rejected parts.

Since it is essential that all the unwanted parts be rejected, the value of b_0 (which defines the position of the cutout apex) must be less than b_u. The value of b_u can be found from either Eq. (3.33) or (3.35), whichever is appropriate.

In practice, this method will result in an effective orienting device but not necessarily the most efficient one. If, for the larger cutout angles, the experi-

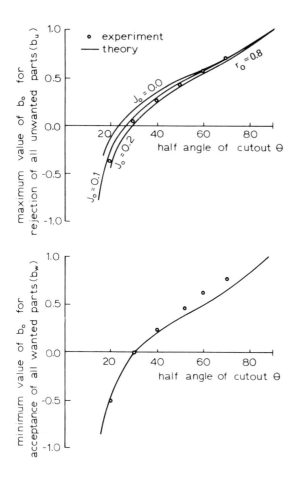

Fig. 3.41 Effect of θ on values of b_u and b_w (from Ref. 8).

mental values of b_w were significantly larger than the theory predicted, the corresponding working ranges $(b_u - b_w)$ would be negative. Thus, when the cutout is designed so that all the unwanted parts are rejected, some of the wanted parts will also be rejected. This reduces the output and can result in low efficiency.

In view of these observations, the recommended procedure would be to choose a value of θ less than 45 deg but large enough to give acceptable deflection of the parts into the bowl and then determine b_u from the appropriate equation. Such an orienting device would have a positive working range when J_0 is less than 0.2 and thus an efficiency of 100%.

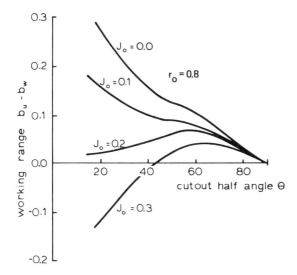

Fig. 3.42 Effect of cutout angle on the working range (from Ref. 8).

3.17 NATURAL RESTING ASPECTS OF PARTS FOR AUTOMATIC HANDLING

In order to analyze the complete system for orienting cup-shaped parts (Fig. 3.31), it was necessary to know the probabilities with which the various orientations would initially occur, and Fig. 3.30 presented the experimental and theoretical results of the distribution of the natural resting aspects for cup-shaped parts.

Figure 3.43 shows another orienting system for a vibratory-bowl feeder. The parts being fed and oriented are cylinders whose L/D ratio is 0.7. The wiper blade is adjusted to reject parts lying on their sides and to allow parts lying on end to pass to the delivery chute. Clearly, the feed rate of oriented parts depends on the rate at which parts encounter the wiper blade and the proportion of these parts that are lying on end. For the case shown in Fig. 3.43, the proportion of parts lying on end is, surprisingly, only about 0.3. This means that only 30% of parts fed to the wiper blade will pass through to the delivery chute.

From the preceding example, it is seen once again that knowledge of the probabilities of the various ways the part will naturally rest (natural resting aspects) is essential in any analysis of the performance of the orienting system. In some cases, it is also necessary to know how a particular natural resting aspect will divide into various orientations on the bowl track. The difference

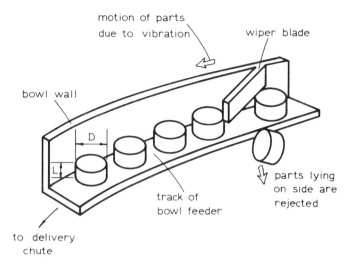

Fig. 3.43 Simple orienting system for a vibratory-bowl feeder ($L/D = 0.7$).

between natural resting aspect and orientation is illustrated in Fig. 3.44. Here, both parts have the same natural resting aspect (that is, lying on their large faces) but have different orientations on the bowl track.

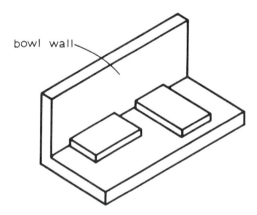

Fig. 3.44 Rectangular parts on a track.

3.17.1 Assumptions

Repeated observations and analyses of the behavior of a variety of parts dropped onto different surfaces have led to the following conclusions [9].

1. Surfaces can be divided into two main categories. First, there is the soft rubber like type of surface (referred to as *soft surfaces*). On impact, a corner of the part digs into the surface, resulting in an impact force having a significant horizontal component in addition to a vertical component. The actions of these force components and the nature of the contact between the part and the surface cause the part to roll across the surface, changing rapidly from one natural resting aspect to another.

The second category of surface is hard and resilient and includes such materials as metals, glass, and hard laminated plastics. With surfaces of this nature (referred to as *hard surfaces*), the corners of the part do not generally dig into the surface, and the horizontal component of the impact force has a negligible effect. After impact, the part does not usually roll across the surface but bounces up and down, overturning repeatedly, but remaining generally in the same area on the surface. Thus, a change of aspect must be brought about by vertical impact forces applied at the edges and corners of the part.

2. The probability that a part will come to rest in a particular natural resting aspect is a function of two factors: (1) the energy barrier tending to prevent a change of aspect, and (2) the amount of energy possessed by the part when it begins to fall into that natural resting aspect.

3. It will be assumed throughout that parts are dropped from a height sufficient to ensure that, after impact, at least one change in natural resting aspect occurs.

3.17.2 Analysis for Soft Surfaces

Consider a square prismatic part initially resting on a flat horizontal surface, with one of its corners, say A, at the origin of a rectangular XY coordinate system, which also lies on the surface. Figure 3.45 shows this part rotated about the X axis through some angle θ small enough so that, if it possessed no kinetic energy, it would fall onto its end. To move onto its side by rolling over corner A, such that the part rotates about a line parallel to the Y axis, the part would require sufficient energy to raise its center of mass from its initial position B to position C when the part is just about to change its aspect.

From the projections in Fig. 3.45, it can be seen that this change of height is equal to

$$BC = [x^2 + \ell^2 \cos^2 (\beta - \theta)]^{1/2} - \ell \cos (\beta - \theta) \qquad (3.37)$$

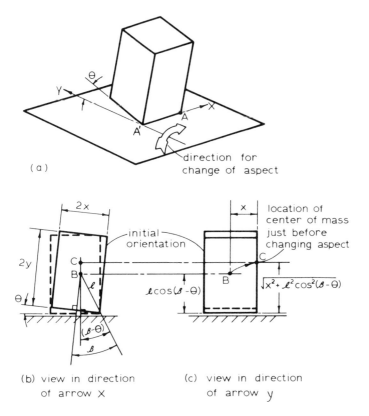

Fig. 3.45 Energy required to change aspect of square prism tilted at an angle θ (from Ref. 9).

where ℓ is given by $(x^2 + y^2)^{1/2}$ and β by arctan (x/y). This expression gives the length of the vertical line BC in Fig. 3.45b, which is the projection of points B and C in Fig. 3.45c onto a line perpendicular to the XY plane. If all such lines (as θ varies from $-\beta$ to $+\beta$) were drawn on the projection in Fig. 3.45b, the shaded area shown in Fig. 3.46a would result. This shaded area represents the total energy barrier for the part resting on end.

The size of the area can be obtained by integrating Eq. (3.37) between the limits $\theta = \pm \beta$. Alternatively, the area can be obtained from Fig. 3.46 by geometry.

The present theory is based on the hypothesis that the probability that a part will come to rest in a particular orientation is proportional to the area of the energy barrier. This is illustrated in Fig. 3.47, which represents the energy barriers for a part with only two resting aspects (say, a regular prism or

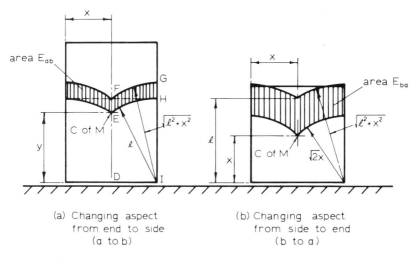

(a) Changing aspect
from end to side
(a to b)

(b) Changing aspect
from side to end
(b to a)

Fig. 3.46 Energy barriers for a square prism (from Ref. 9).

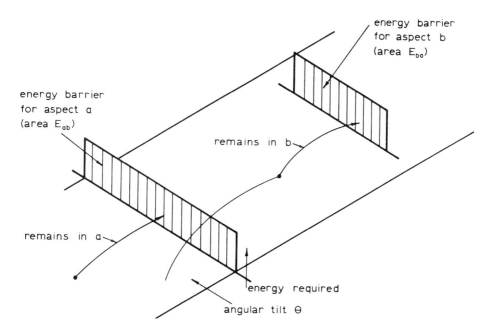

Fig. 3.47 Representation of energy barriers showing that the number of parts remaining in an aspect will be proportional to the energy barrier for that aspect (from Ref. 9).

cylinder that can rest only on its side or on end). As the part tumbles end over end, it will encounter, on each change of aspect, the appropriate energy barrier. If it is assumed that as the part passes through each aspect, the probability that it possesses a given amount of kinetic energy is independent of the angle of tilt, then the probability that the part will not surmount the energy barrier will be proportional to the area of the barrier.

Thus, referring to Fig. 3.47, the number of parts changing aspect from a to b that will be stopped by the first energy barrier will be proportional to the area of the energy barrier E_{ab}. Similarly, the number stopped by the second energy barrier will be proportional to E_{ba}. This situation will continue until the part comes to rest.

Thus, in the end, the number of parts N_a and N_b remaining in aspects a and b will be given, respectively, by

$$N_a = E_{ab} \tag{3.38}$$

$$N_b = E_{ba}$$

Hence, returning to the example in Fig. 3.46, by geometry,

$$E_{ab} = 2(\text{area enclosed by points } EFGH) \tag{3.39a}$$

or

$$E_{ab} = 2(A_{DFI} + A_{IFG} - A_{DEI} - A_{IEH}) \tag{3.39b}$$

where A_{DFI} refers to the area enclosed by points D, F, I; A_{IFG} refers to the area enclosed by points I, F, G; and so on.

Thus, it can be shown that

$$E_{ab} = x^2 \left[\alpha_2 p^2 + q - \alpha_1 q^2 - \frac{y}{x} \right] \tag{3.40}$$

where

$$p = \frac{(\ell^2 + x^2)^{1/2}}{x} = \left[2 + \left(\frac{y}{x} \right)^2 \right]^{1/2}$$

$$q = \frac{\ell}{x} = \left[1 + \left(\frac{y}{x} \right)^2 \right]^{1/2}$$

$$\alpha_1 = \arcsin \left[\frac{1}{q} \right]$$

$$\alpha_2 = \arcsin \left[\frac{1}{p} \right]$$

$$E_{ba} = x^2 \left[\alpha x^2 p^2 + q - \frac{\pi}{2} - 1 \right] \qquad (3.41)$$

Since the part has two ends and four sides, the probability for aspect a (on end) is

$$P_a = \frac{2E_{ab}}{2E_{ab} + 4E_{ba}} \qquad (3.42)$$

and the probability for aspect b (on the side) is

$$P_b = 1 - P_a \qquad (3.43)$$

Calculation of the values of E_{ab} and E_{ba} from Eqs. (3.40) and (3.41), and substitution in Eqs. (3.42) and (3.43) give the curve shown in Fig. 3.48. Also shown in this figure are experimental results for steel parts dropped onto a rubber-coated (soft) surface.

The energy barrier for a solid cylindrical part is obtained in a similar manner. Figure 3.49a shows such a part that was initially resting on end on a flat horizontal surface but that has been rotated about the X axis through a small angle θ so that it would fall onto its end if it were released. To move the

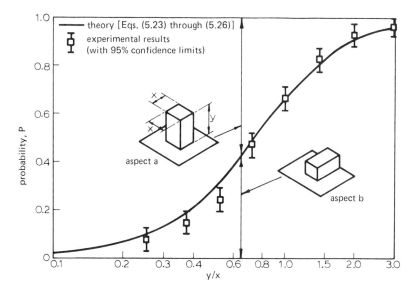

Fig. 3.48 Probabilities of natural resting aspects for square prisms dropped onto a soft surface (from Ref. 9).

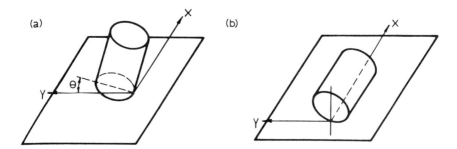

Fig. 3.49 (a) Solid cylindrical part tilted through an angle θ and (b) solid cylindrical part lying on its side.

cylinder onto its side by now rotating it about the Y axis causes the center of mass to rise further, as shown by the energy barrier in Fig. 3.50a.

If the part were first lying on its side, as shown in Fig. 3.49b, and then rolled on the horizontal surface, the distance from the center of mass to the surface would remain constant. If the part were now rotated onto one end about one of its edges, the energy barrier becomes that shown in Fig. 3.50b.

In these cases, the expressions for the areas of the energy barriers are rather simpler than in the case of a square prism, and

$$E_{ab} = \ell^2[(2 - \cos \alpha) \sin \alpha - \alpha] \tag{3.44}$$

$$E_{ba} = \ell^2(1 - \sin \alpha)\pi \sin \alpha \tag{3.45}$$

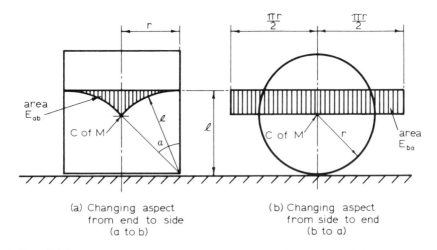

Fig. 3.50 Energy barriers for a cylinder on a soft surface (from Ref. 9).

where $\alpha = \cot(L/D)$ and L is the length and D the diameter of the part. Hence,

$$P_a = \frac{E_{ab}}{E_{ab} + E_{ba}} \tag{3.46}$$

$$= \frac{(2 - \cos\alpha) - (\alpha/\sin\alpha)}{(2 - \cos\alpha) - (\alpha/\sin\alpha) + \pi(1 - \sin\alpha)}$$

and

$$P_b = 1 - P_a \tag{3.47}$$

Rather surprisingly, it is found that, if the results for a square prism (Fig. 3.48) are plotted in terms of the L/D ratio of the circumscribed cylinder, they can be very closely approximated by the results in Fig. 3.51 for a solid cylinder. Indeed, application of the theory to prisms of any regular cross section (triangular, square, pentagonal, hexagonal, etc.), together with experiments, shows that Eqs. (3.44–3.47) apply to any solid regular prism, where L/D is defined as the length-to-diameter ratio of the circumscribed cylinder. This is illustrated in Fig. 3.52.

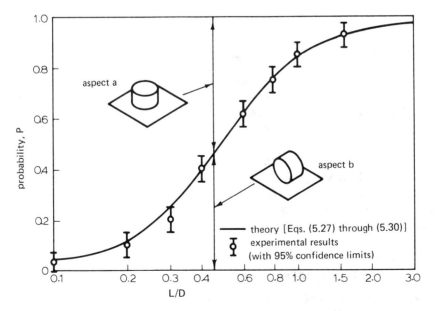

Fig. 3.51 Probabilities of natural resting aspects for cylinders dropped onto a soft surface (from Ref. 9).

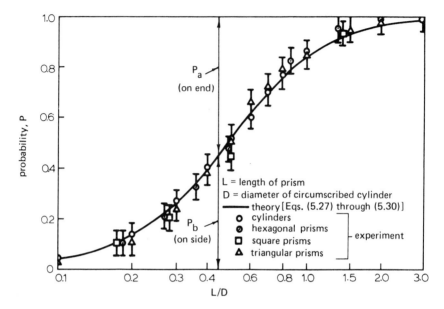

Fig. 3.52 Probabilities of natural resting aspects for cylinders and various regular prisms dropped onto a soft surface (from Ref. 9).

3.17.3 Analysis for Hard Surfaces

For hard surfaces, the approach described in Boothroyd et al. [10] can be used in conjunction with the new energy-barrier technique described above. In this case, it is assumed that an extra amount of energy is required to change a part aspect because energy can be provided only by a vertical impact force applied at one of the edges or corners of the part. In addition, allowance must be made for the effect of this additional energy, which tends to keep the part from remaining in the next aspect.

Again, as with the soft-surface solution and experiments, it was found that the results for a solid cylinder could be applied to any prism of regular cross section with three or more sides that could be enclosed within the cylinder.

3.17.4 Analysis for Cylinders and Prisms with Displaced Centers of Mass

If the center of mass of a cylinder is displaced along its axis from the midpoint, the probabilities that the cylinder will come to rest on either of its ends become different. Based on experimental and theoretical work, it was found that the approach described above for both hard and soft surfaces could be applied equally successfully to this problem.

3.17.5 Summary of Results

Figure 3.53 presents a complete summary of the results of this work. It can be seen that the probabilities for the natural resting aspects of all prisms of regular cross section can be presented in only two graphs (one for a soft surface and one for a hard surface).

3.18 ANALYSIS OF A TYPICAL ORIENTING SYSTEM

In general, an orienting system for a vibratory-bowl feeder consists of one or more orienting devices arranged in series along the bowl track. These devices are usually located near the outlet of the feeder along a horizontal portion of the track. Because the track section is level, the parts can travel at a conveying velocity that is greater than the velocity of the parts on the preceding incline, thus enabling parts to separate and eliminating interference between adjacent parts as they pass the various devices. Such a system is shown in Fig. 3.54 for the feeding and orienting of right rectangular prisms. The six orientations for these prisms are also described in this figure.

The first device is a wiper blade, and it rejects orientations *c, d, e,* and *f* back into the bowl. It also serves to remove the secondary layers of parts, where one part rests on another instead of on the track. The output of this device is either orientation *a* or *b*.

The narrow track is next, and it rejects orientation *b* back into the bowl, leaving only orientation *a*. The riser turns orientation *a* into orientation *c*, which is the output orientation of the system.

The matrices for these devices and systems are:

$$
\begin{array}{cccc}
\text{Wiper blade} & \text{Narrow track} & \text{Riser} & \text{System} \\[4pt]
\begin{array}{c}
 \\
a \\ b \\ c \\ d \\ e \\ f
\end{array}
\begin{array}{c}
\begin{matrix} a & b \end{matrix} \\
\begin{bmatrix} 1 & 0 \\ 0 & 1 \\ 0 & 0 \\ 0 & 0 \\ 0 & 0 \\ 0 & 0 \end{bmatrix}
\end{array}
&
\begin{array}{c}
 \\
a \\ b
\end{array}
\begin{array}{c}
\begin{matrix} a \end{matrix} \\
\begin{bmatrix} 1 \\ 0 \end{bmatrix}
\end{array}
&
\begin{array}{c}
 \\
a
\end{array}
\begin{array}{c}
\begin{matrix} c \end{matrix} \\
\begin{bmatrix} 1 \end{bmatrix}
\end{array}
\;=\;
&
\begin{array}{c}
 \\
a \\ b \\ c \\ d \\ e \\ f
\end{array}
\begin{array}{c}
\begin{matrix} c \end{matrix} \\
\begin{bmatrix} 1 \\ 0 \\ 0 \\ 0 \\ 0 \\ 0 \end{bmatrix}
\end{array}
\end{array}
$$

Rectangular prisms, when tossed on a horizontal surface, can come to rest on one of three faces. These positions are the three natural resting aspects for these parts. The probabilities that a prism will come to rest in each of these three aspects are shown in Fig. 3.55. The parts used in this particular system

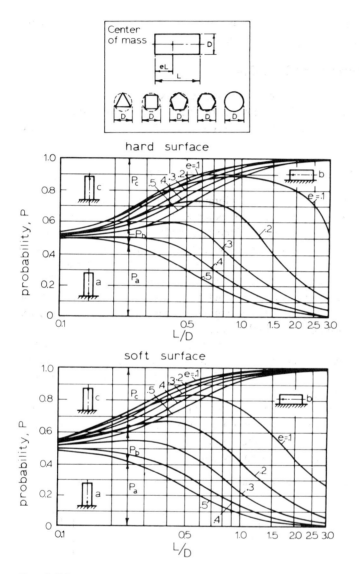

Fig. 3.53 Probabilities of natural resting aspects for prisms of regular cross section; L is the length of prism and D the diameter of circumscribed cylinder (from Ref. 9).

are $45 \times 30 \times 3$ mm. The values c/a and c/b are 0.07 and 0.10, respectively. According to Fig. 3.55, virtually all these parts will rest on their largest face when tossed onto a hard horizontal surface, such as the bottom of a bowl

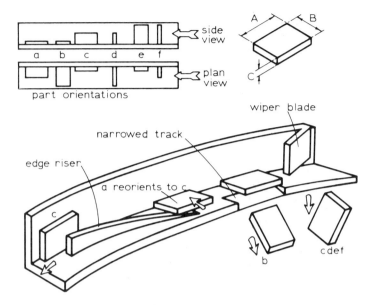

Fig. 3.54 Orienting system for right rectangular prisms.

feeder. However, within this one natural resting aspect, there are two orientations (a and b). Parts in the bottom of the bowl tend to rotate into one of these orientations before they travel up the inclined track. For aluminum parts on a steel track, the coefficient of friction μ is 0.4. Thus, from Ho and Boothroyd [12], or Fig. 3.56, it is seen that, for these parts, the partition ratio $R = P_a/(P_a + P_b)$ is 0.63 and, therefore, the probabilities are given by

$$P_a = R(P_a + P_b) = 0.63, \qquad P_b = 0.37$$

The initial distribution matrix (IDM) showing the probable distribution of the various orientations is, therefore,

$$\begin{matrix} a & b & c & d & e & f \\ [0.63 & 0.37 & 0 & 0 & 0 & 0] \end{matrix}$$

Thus, the efficiency η of this system, which is given by the product of the IDM and the system matrix, is 63%.

The average length of a part entering this system is the product of the IDM and a matrix of the lengths of the corresponding orientations in the conveying direction. For these parts, the average part length $\bar{\ell}$ is

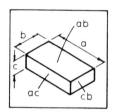

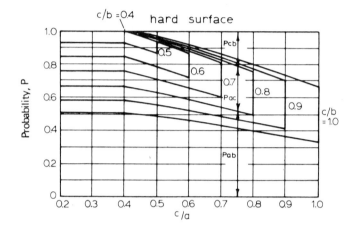

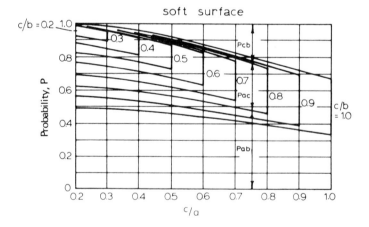

Fig. 3.55 Probabilities of natural resting aspects for right rectangular prisms (from Ref. 9).

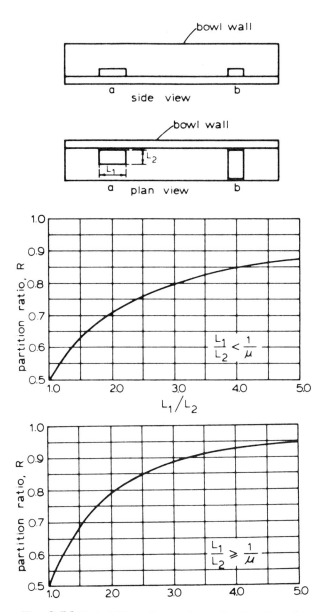

Fig. 3.56 Probabilities of orientations of rectangular prisms on a bowl-feeder track. P_a is the probability of orientation a; P_b is the probability of orientation b; partition ratio $R = P_a/(P_a + P_b)$; μ is the coefficient of friction between parts and bowl track; and L_1 and L_2 are the length of long and short sides, respectively, of the part-track interface ($L_1 \geqslant L_2$) (from Ref. 12).

$$[0.63\ 0.37\ 0\ 0\ 0\ 0]\begin{bmatrix} 45 \\ 30 \\ 45 \\ 3 \\ 30 \\ 3 \end{bmatrix} = 39\,\text{mm}$$

The feed rate F can be found from

$$F = \eta\frac{v}{\ell} \qquad (3.48)$$

where v is the conveying velocity of the parts on the inclined section of the track when adjacent parts are touching. The feed rate of any system can be found in a similar manner, using the initial distribution matrix, system matrix, length matrix, and Eq. (3.48). The feed rate can also be determined from

$$F = vE/A \qquad (3.49)$$

where E is known as the modified efficiency of the orienting system and is given by

$$E = \eta A/\bar{\ell} \qquad (3.50)$$

3.18.1 Design of Orienting Devices

To achieve the calculated 63% efficiency, the orienting devices that make up the system must be properly designed. Design data and performance curves for orienting devices used in vibratory-bowl feeders are provided in a *Handbook of Feeding and Orienting Techniques for Small Parts* [11].

For the wiper blade, which rejects orientations c, d, e, and f (Fig. 3.54), the angle between the wiper blade and the bowl wall θ is set to avoid the jamming action produced by overlapping parts, as shown in Fig. 3.57. The smallest jamming angle β_w for these parts is arctan(3/45) or 3.8 deg (0.07 rad), and the maximum value of θ_w, from Fig. 3.57 is 18 deg (0.31 rad). The height of the wiper blade should be sufficient to remove a secondary layer of parts or 5 mm.

The narrow-track device rejects orientation b but allows orientation a to pass. From Fig. 3.58 and a conveying velocity of 100 mm/s, the corresponding values for the dimensionless track width b_t/w are 1.2 and 1.45, respectively. Thus, in millimeters,

$$(1.2)(22.5) > b_t > (1.45)(15)$$

or

$$27 > b_t > 22$$

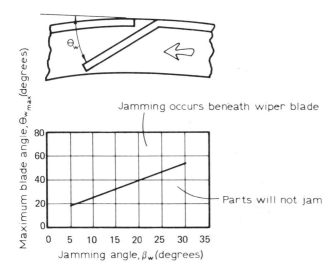

Fig. 3.57 Wiper blade design (adapted from Ref. 11).

The narrow track should be 23 mm wide and 67 mm long.

The edge-riser orienting device turns orientation a into orientation c. The design information for this device is presented in Fig. 3.59. For a 6-deg (0.10-rad) riser angle and parts $45 \times 30 \times 3$ mm, B/C equals 10, and the ramp length is 300 mm. Other riser angles and lengths will also produce satisfactory results.

3.19 OUT-OF-BOWL TOOLING

A further type of orienting device is that which is situated between the feeder and the workhead. Such devices are usually of the active type because orientation by rejection is not often practicable. Figure 3.60 illustrates a device described by Tipping [13] in which the position of the center of gravity of a part is utilized. In this example, the cup-shaped part is pushed onto a bridge, and the weight of the part acting through the center of gravity pulls the part

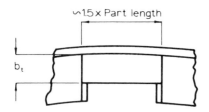

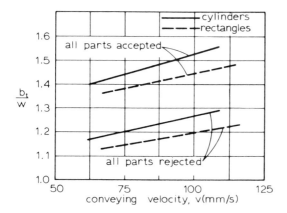

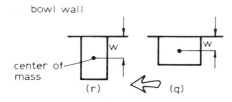

Fig. 3.58 Narrow-track design (adapted from Ref. 11).

down nose first into the delivery chute. In Fig. 3.61, the same part is reoriented using a different principle. With this method, if the part passes nose first down the delivery tube, it is deflected directly into the delivery chute and maintains its original orientation. A part fed open end first will be reoriented by the pin located in the wall of the device.

The device illustrated in Fig. 3.62 is known as a selector and employs a principle that has been applied successfully to reorient a wide variety of parts [14]. The selector consists of a stationary container in which a wheel with radial slots is mounted. The wheel is driven by an indexing mechanism to

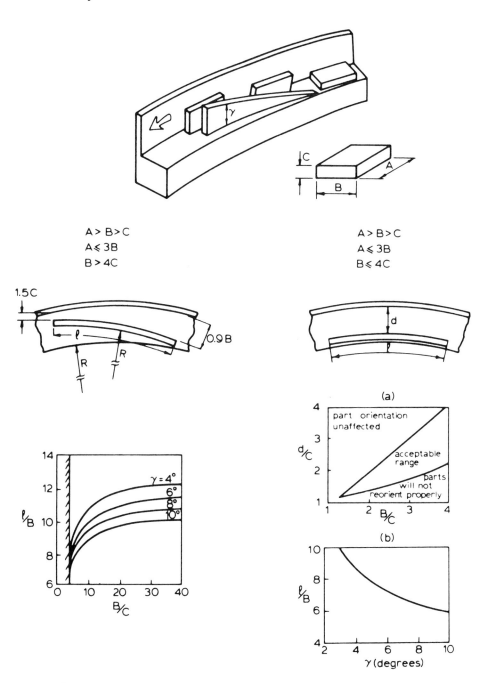

Fig. 3.59 Design data for the edge-riser orienting device (adapted from Ref. 11).

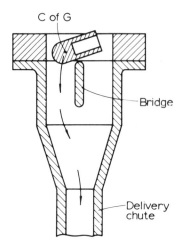

Fig. 3.60 Reorientation of cup-shaped part (C of G is center of gravity) (from Ref. 13).

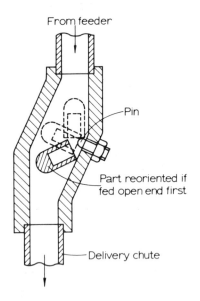

Fig. 3.61 Reorientation of cup-shaped part (from Ref. 13).

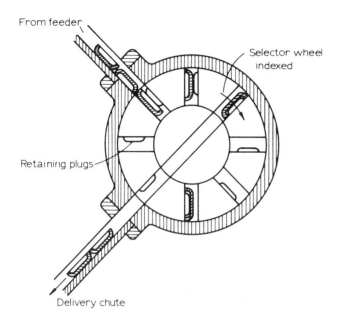

Fig. 3.62 Reorientation of shallow-drawn parts (from Ref. 14).

ensure that the slots always align with the chutes. In the design illustrated, the shallow-drawn parts may enter the slot in the selector wheel in either of two attitudes. After two indexes, the parts are aligned with the delivery chute and those that now lie open end upward slide out of the selector and into the delivery chute. Those that lie open end downward are retained by a plug in the slot. After a further four indexes, the slot is again aligned with the delivery chute.

REFERENCES

1. Redford, A.H., and Boothroyd, G., "Vibratory Feeding," Proc. I Mech. Eng., Vol. 182, Part 1, No. 6, 1967–1968, p. 135.
2. Redford, A.H., "Vibratory Conveyors," Ph.D. Thesis, Salford University, Salford, England, 1966.
3. Boothroyd, G., Poli, C.R., and Murch, L.E., "Feeding and Orienting Techniques for Small Parts," SME Technical Paper AD72-763, 1975.
4. Jimbo, Y., Yokoyama, Y., and Okabe, S., "Vibratory Conveying," *Bulletin of the Japan Society of Precision Engineering*, Vol. 4, No. 3, 1970, pp. 59–64.

5. Yokoyama, Y., Okabe, S., Shiozawa, A., and Watanabe, M., "The Balanced Vibratory Feeder and Its Application for Automated Assembly System," *Proceedings 4th International Conference on Production Engineering*, Tokyo, 1980, pp. 907–912.

6. Murch, L.E., and Boothroyd, G., "Design of Orienting Systems for Vibratory Bowl Feeders," SME Technical Paper FC72–235, 1972.

7. Murch, L.E., and Boothroyd, G., "Predicting Efficiency of Parts Orienting Systems," *Automation*, Vol. 18, Feb. 1971.

8. Boothroyd, G., and Murch, L.E., "Performance of an Orienting Device Employed in Vibratory Bowl Feeders," *Transactions of the ASME, Journal of Engineering for Industry*, Aug. 1970.

9. Boothroyd, G., and Ho, C., "Natural Resting Aspects of Parts for Automatic Handling," *Transactions of the ASME, Journal of Engineering for Industry*, Vol. 99, May 1977, pp. 314–317.

10. Boothroyd, G., Redford, A.H., Poli, C., and Murch, L.E., "Statistical Distribution of Natural Resting Aspects of Parts for Automatic Handling," *Manufacturing Engineering Transactions*, Vol. 1, 1972.

11. Boothroyd, G., Poli, C., and Murch, L.E., *Handbook of Feeding and Orienting Techniques for Small Parts*, University of Massachusetts, Amherst, Mass., 1977.

12. Ho, C., and Boothroyd, G., "Orientation of Parts on the Track of A Vibratory Feeder," *Proceedings of the Fifth North American Metalworking Research Conference*, SME, Dearborn, Michigan, 1977, p. 363.

13. Tipping, W.V., "Mechanized Assembly Machines, 9: Orientation and Selection," *Machine Design Engineering*, Feb. 1966, p. 36.

14. Hopper Feeds as an Aid to Automation, *Machinery's Yellowback No. 39*, Machinery Publishing, Brighton, England.

4

Automatic Feeding and Orienting — Mechanical Feeders

Although the vibratory-bowl feeder is the most widely employed and most versatile parts-feeding device, many other types of parts feeders are available. Usually, these are suitable only for feeding certain basic types of component parts but, for feeding these parts, better results may be obtained for a smaller capital outlay with feeders other than the vibratory type.

One point that must be borne in mind when we consider parts feeders is that, in automatic assembly, the output of parts from the feeder is always restricted by the machine being fed. The machine generally uses parts at a strictly uniform rate, which may be referred to as the *machine rate*. In the design and testing of parts feeders, it is often convenient to observe the feed rate when the feeder is not connected to a machine, that is, when no restriction is applied to the output of the feeder. The feed rate under these circumstances will be referred to as the *unrestricted feed rate*. Clearly, in practice, the mean unrestricted feed rate must not fall below the machine rate.

Certain other general requirements of parts feeders may be summarized as follows: the unrestricted feed rate should not vary widely because this simply means that when the feeder is connected to a machine, the parts are continuously recirculated within the feeder for much of the time. This causes excessive wear and may eventually damage the parts. This undesirable characteristic often occurs in parts feeders in which the feed rate is sensitive to changes in the quantity of parts present in the feeder and will be referred to as the *load sensitivity* of the feeder.

With parts feeders suitable for automatic machines, it is necessary that all the parts be presented to the machine in the same orientation; that is, they must be fed correctly oriented. Some feeders are able to feed and orient many types of parts, whereas others are able to handle only a very limited range of part shapes.

Undoubtedly, the reliability of a parts feeder is one of its most important characteristics. Parts feeders should be designed so that the possibility of parts jamming in the feeder, or in its orienting devices, is minimized or eliminated.

It is sometimes suggested that parts feeders can also act as inspection devices. It is possible to design certain parts feeders so that misshapen parts, swarf, and so on, will not be fed to the machine but will be rejected by the device fitted to the feeder. This can be an important feature because defective parts or foreign matter, if fed to the machine, will probably cause a breakdown and may stop the whole production line.

Some parts feeders are noisy in operation, and some tend to damage certain types of parts. Obviously, both these aspects of parts feeding must be considered when studying the possible alternatives for a particular application.

Parts feeders can generally be classified into the following: reciprocating feeders; rotary feeders; belt feeders; and vibratory feeders, which were discussed in Chapter 3. A selection of the more common feeding devices within each of these groups will now be described and discussed.

4.1 RECIPROCATING-TUBE HOPPER FEEDER

A reciprocating-tube hopper is illustrated in Fig. 4.1 and consists of a conical hopper with a hole in the center, through which a delivery tube passes. Relative vertical motion between the hopper and the tube is achieved by reciprocating either the tube or the hopper. During the period when the top of the tube is below the level of parts, some parts will fall into the delivery tube. It is usual to machine the top of the tube at an angle so that a part resting across the opening will fall clear and not block the opening as the tube is pushed upward through the mass of parts. Care must be taken in choosing the angle of the conical hopper because, if the angle is too small, parts may jam between the tube and the hopper.

Figure 4.2 shows the forces acting on a cylindrical part jammed in this way when the tube is moving downward relative to the hopper. The force W acting vertically downward represents the weight of the part, together with any additional force that may be present as a result of parts resting on top of the one

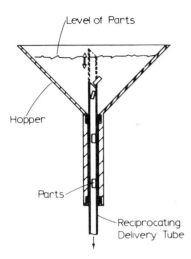

Fig. 4.1 Reciprocating-tube hopper.

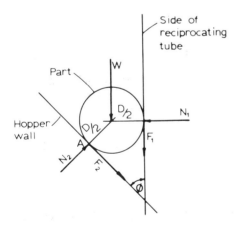

Fig. 4.2 Forces acting on a part jammed between the hopper wall and the tube.

shown. Resolving forces vertically and horizontally and taking moments about A gives

$$F_1 + W + F_2 \cos \phi = N_2 \sin \phi \qquad (4.1)$$

$$N_1 = N_2 \cos \phi + F_2 \sin \phi \qquad (4.2)$$

$$F_1(1 + \cos \phi)\, D/2 + W(D/2) \cos \phi = N_1(D/2) \sin \phi \qquad (4.3)$$

where ϕ is the hopper wall angle and D the diameter of the part.

Eliminating W from Eqs. (4.1) and (4.3) gives, after rearrangement,

$$N_1 \sin \phi = F_1 (1 + \cos \phi) + \cos \phi$$

$$\times (N_2 \sin \phi - F_2 \cos \phi - F_1) \qquad (4.4)$$

The maximum value of F_2 is given by $\mu_s N_2$ (where μ_s is the coefficient of static friction) and, thus, writing $F_2 = \mu_s N_2$ in Eqs. (4.2) and (4.4) and eliminating N_2 give

$$\frac{F_1}{N_1} = \frac{\mu_s}{\cos \phi + \mu_s \sin \phi} \qquad (4.5)$$

For the tube to slide, $F_1/N_1 > \mu_s$, and, therefore, from Eq. (4.5),

$$\frac{1}{\cos \phi + \mu_s \sin \phi} > 1 \qquad (4.6)$$

Expression (4.6) indicates that the value of ϕ should be as large as possible to prevent jamming when μ_s is large. However, when $\mu_s < \cot \phi$, the parts cannot slide down the hopper wall. The best compromise is probably given by writing the limiting conditions

$$\mu_s = \cot \phi \qquad (4.7)$$

$$\cos \phi + \mu_s \sin \phi = 1 \qquad (4.8)$$

Combining Eqs. (4.7) and (4.8) gives $\phi = 60$ deg and, on substitution of this value in expression (4.6), it is found that, to prevent jamming under these conditions, the coefficient of friction μ_s must be less than 0.577. Since this value is greater than that expected in practice, it may be concluded that, with a hopper angle of 45 deg, the possibility of jamming will generally be avoided if the coefficient of friction is less than 0.414.

4.1.1 General Features

The optimum hopper load is that which fills half the volume of the hopper, and the delivery tube should rise just above the maximum level of parts in the hopper. The inside silhouette of the delivery tube must be designed to accept only correctly oriented parts one at a time. The linear velocity of the delivery tube should be no greater than 0.6 m/s.

4.1.2 Specific Applications

Figure 4.3 shows some specific results [1] for the feeding of cylinders. From the figure it is possible to estimate the feed rates obtainable for a range of cylindrical parts.

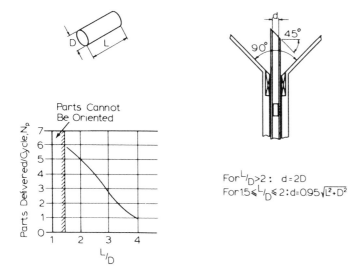

$$\text{For} {}^L\!/_D > 2: \quad d = 2D$$
$$\text{For} 1.5 \leqslant {}^L\!/_D \leqslant 2 : d = 0.95\sqrt{L^2 + D^2}$$

Fig. 4.3 Performances of a reciprocating-tube hopper when feeding cylindrical parts (adapted from Ref. 1).

4.2 CENTERBOARD HOPPER FEEDER

Figure 4.4 shows a typical centerboard hopper feeder. Basically, this consists of a hopper, in which the parts are placed at random, and a blade with a shaped track along its upper edge, which is periodically pushed upward through the mass of parts. The blade will thus catch a few parts on its track during each cycle; when the blade is in its highest position (as shown in the figure), it is aligned with a chute, and the parts will slide down the track and into the chute. The centerboard hopper illustrated is suitable for feeding cylindrical parts.

4.2.1 Maximum Track Inclination

One of the important parameters in a centerboard hopper design is the angle of inclination of the track when the blade is in its highest position (θ_m in Fig. 4.4). It is assumed, for the purposes of the following analysis, that the cam drive is arranged so that the blade is lifted rapidly to its highest position, allowed to dwell for a period while the parts slide into the chute, and then rapidly returned to its lowest position when the track is horizontal and aligned with the bottom of the hopper.

Clearly, there is a limit on the deceleration of the blade on its upward stroke; otherwise, the parts leave the track and are thrown clear of the feeder.

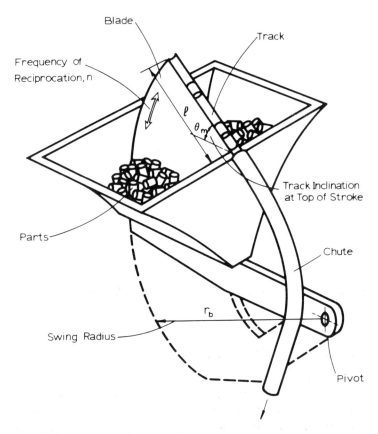

Fig. 4.4 Centerboard hopper feeder.

Thus, for a given deceleration, an increase in the angle θ_m increases the time taken for the blade to complete its upward motion. However, with larger values of θ_m, the time taken for the parts to slide off the track is less and, when θ_m is chosen to give maximum frequency of reciprocation and hence maximum feed rate, a compromise must be sought.

The tendency for a part to leave the track during the upward motion of the blade is greatest at the end of the track farthest away from the pivot. The forces acting on a part in this position are shown in Fig. 4.5 and, from the figure, the condition for the reaction between the part and the track to become zero is given by

$$m_p \alpha \left[r_b - \frac{L}{2} \right] = -m_p g \cos \theta_m \qquad (4.9)$$

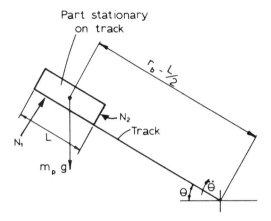

Fig. 4.5 Forces acting on a part during the upward motion of the blade.

where L is the length of the part, m_p the mass of the part, r_b the radius from the pivot to the upper end of the track, θ_m the maximum angle between the track and the horizontal, and α the angular acceleration of the track.

Thus, the maximum angular deceleration of the blade is approximately given by

$$-\alpha = \frac{g}{r_b} \cos \theta_m \tag{4.10}$$

if L is small compared with r_b.

For simplicity, it is now assumed that, during the period of the upward motion of the blade, the drive to the blade is designed to give (1) a constant acceleration of $(g \cos \theta_m)/r_b$ followed by (2) a constant deceleration of $(g \cos \theta_m)/r_b$. Under these conditions, the total time t_1 taken to lift the blade so that the track is inclined at an angle θ_m to the horizontal is given by

$$t_1^2 = \frac{4r_b\theta_m}{g \cos \theta_m} \tag{4.11}$$

It is now assumed that when the blade is in its highest position, it dwells for a period t_2, just long enough to allow the parts to slide down the track. This is given, in the worst case, by the time taken for one part to slide the whole length of the track. The forces acting on a part under these circumstances are shown in Fig. 4.6, and resolving in a direction parallel to the track gives

$$m_p a = m_p g \sin \theta_m - \mu_d m_p g \cos \theta_m \tag{4.12}$$

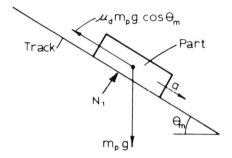

Fig. 4.6 Forces acting on a part as it slides down a track.

where a is the linear acceleration of the part down the track, and μ_d is the coefficient of dynamic friction between the part and the track. The minimum dwell period t_2 is now given by

$$t_2^2 = \frac{2\ell}{g(\sin\theta_m - \mu_d \cos\theta_m)} \tag{4.13}$$

where ℓ is the total length of the track.

If the time taken to return the blade to its lowest position is now assumed to be the same as the time for the up stroke, then the total period t_f of the feeder cycle is given by

$$t_f = 2t_1 + t_2 = 2 \left[\frac{4r_b\theta m}{g\cos\theta_m} \right]^{1/2}$$

$$+ \left[\frac{2\ell}{g(\sin\theta_m - \mu_d\cos\theta_m)} \right]^{1/2} \tag{4.14}$$

Equation (4.14) consists of two terms; one that increases as θ_m is increased and one that decreases as θ_m is increased. An optimum value of θ_m always exists that gives the minimum period t_f and, hence, a maximum theoretical feed rate. It can be shown mathematically that this optimum value of θ_m is a function only of μ_d and the ratio r_b/ℓ. However, the resulting expression is unmanageable, but the curve shown in Fig. 4.7a gives the solution for a practical value of r_b/ℓ of 2.0. For example, with a coefficient of dynamic friction of 0.4, the optimum track angle would be approximately 36 deg.

Figure 4.7b shows how the maximum frequency of the blade cycle $n_{\max}$ (given by $1/t_f$) varies as the coefficient of friction between part and track is changed and when the ratio r_b/ℓ is 2.0. It can be seen that, for large values of μ_d in the range 0.4–0.8, the maximum blade frequency varies by only 10–

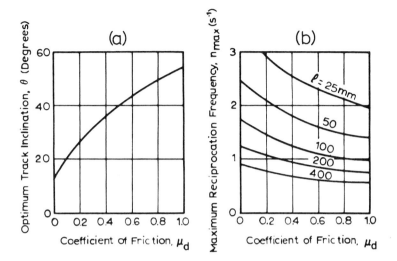

Fig. 4.7 Characteristics of centerboard hopper ($r_b/l = 2.0$).

15%. The maximum blade frequency is more sensitive to changes in the length ℓ of the track and, for longer tracks, the frequency is lower. However, it should be remembered that, for a given size of a part, a longer track, on average, picks up a greater number of parts per cycle and, hence, the mean feed rate may increase.

The maximum number of parts that may be selected during each cycle is given by ℓ/L. In practice, the average number selected is less than this and, if E is taken to be the efficiency of a particular design, the average number of parts fed during each cycle is given by $E\ell/L$ and the mean feed rate F of the hopper feeder is given by

$$F = \frac{nE\ell}{L} \tag{4.15}$$

where the blade frequency n is given by

$$n = \frac{1}{t_f} \tag{4.16}$$

In practice, the values of the efficiency E must be obtained from experiments.

4.2.2 Load Sensitivity and Efficiency

For a centerboard hopper feeder working at a constant frequency, any variation in feed rate as the hopper gradually empties will be due to changes in the

efficiency E. This has been defined as the ratio between the average number of
parts selected during one cycle and the maximum number that can be selected.
Figures 4.8 and 4.9 show the results of tests on an experimental feeder where
$\ell/L = 6$, $r_b/\ell = 2$, and $\theta_m = 54$ deg. The lower curve in Fig. 4.8 shows how
the efficiency E varied as the hopper gradually emptied, and it is interesting to
note that a rapid increase in E occurs when less than 100 parts remain in the
hopper. The upper curve in Fig. 4.8 shows that this high efficiency can be
maintained for all hopper loadings if a baffle is placed on one side of the
hopper blade. The baffle would therefore appear to affect the orientation of
those parts likely to be selected by the blade. Clearly, the load-sensitivity
characteristics obtained with this latter design approach the ideal situation very
closely. Each of the experimental points in Fig. 4.8 represents the average of

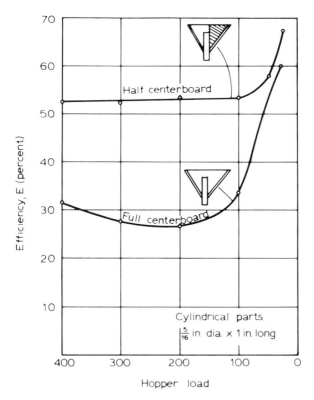

Fig. 4.8 Load sensitivity of a centerboard hopper.

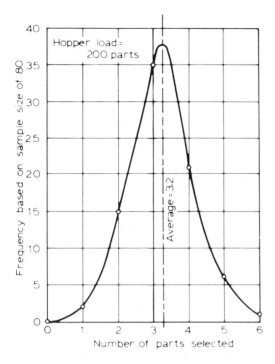

Fig. 4.9 Frequency of part selection of a centerboard hopper.

80 results. In Fig. 4.9, one set of 80 results (for a hopper load of 200 parts) is plotted in the form of a histogram and exhibits the familiar normal distribution. The average number of parts selected in this case was 3.2 and, since the maximum number of parts that could be selected by the blade was 6, this represents an efficiency of 0.53.

Figure 4.10 illustrates a specific application for the centerboard hopper feeder. In the figure, the effect of part proportions on the efficiency E is presented, which allows an estimate of the feed rate to be made.

A final design consideration is the inclination of the sloping sides of the hopper. If the inclination is too great, there is a possibility that parts will jam between the hopper wall and the blade when the blade is moving downward. This situation is identical to that of the reciprocating-tube hopper, and the analysis indicated that the included angle between the hopper wall and the blade should be 60 deg.

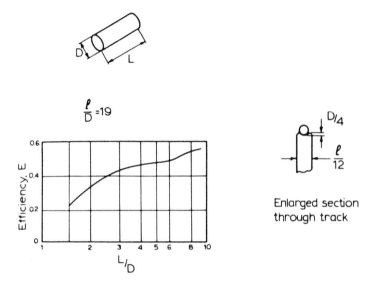

Fig. 4.10 Performance of a centerboard hopper when feeding cylinders (ℓ = length of track) (adapted from Ref. 1).

4.3 RECIPROCATING-FORK HOPPER FEEDER

A reciprocating-fork hopper, shown in Fig. 4.11, is suitable only for feeding headed parts. It consists of a shallow cylindrical bowl that rotates about an axis inclined at approximately 10 deg to the vertical and a fork that reciprocates in the vertical plane about point A. In its lowest position, the fork is horizontal, and rotation of the bowl causes parts to be caught in the fork. The fork then lifts a few parts by their heads to a height sufficient to cause the parts to slide off the fork and into the delivery chute. The analysis for the maximum fork inclination and the maximum rate of reciprocation would be similar to that presented above for a centerboard hopper. The number of parts selected by the fork per cycle would be obtained by experiment.

4.4 EXTERNAL GATE HOPPER FEEDER

An external gate hopper (Fig. 4.12) basically consists of a rotating cylinder having slots in its wall so that the cylindrical parts, if oriented correctly, can nest against the wall of the stationary outer sleeve. At some point, as the cylinder rotates, the slots pass over a gate in the outer sleeve, which allows the parts to drop one by one into the delivery chute. The tumbling action caused by rotation of the cylinder provides repeated opportunities for parts to fall into the slots and, subsequently, to pass through the external gate into the chute.

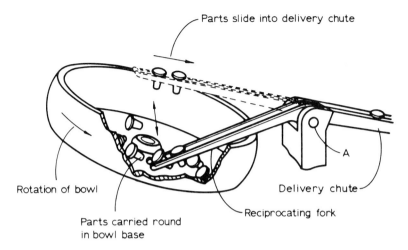

Fig. 4.11 Reciprocating-fork hopper.

4.4.1 Feed Rate

Figure 4.13a shows an enlarged cross section of the slot and part just before the part falls through the gate. In the following analysis, an equation is developed for the maximum peripheral velocity of the inner cylinder that permits feeding to occur. Clearly, if the velocity is too high, the part will pass over the gate. At some limiting velocity v, the part will neither fall through the gate nor pass over it but will become jammed between corners B and C of the slot and gate as shown in Fig. 4.13b. With any velocity below v, the part will drop through the gate as shown in Fig. 4.13c. The position shown in Fig. 4.13a represents the point at which the part starts to fall. In Fig. 4.13b, the part has moved from this position a horizontal distance

$$D - 0.5(D^2 - h_g^2)^{1/2}$$

which, at a velocity v, represents a time interval of

$$\frac{D - 0.5(D^2 - h_g^2)^{1/2}}{v}$$

During this time, the part has fallen a distance $(D/2 - h_g/2)$ and, if it is assumed that the part has fallen freely, the time taken is given by $[2(D/2 - h_g/2)/g]^{1/2}$. Thus, when these times are equated, the limiting velocity is given by

$$\frac{D - 0.5(D^2 - h_g^2)^{1/2}}{v} = \left[\frac{D - h_g}{g}\right]^{1/2} \tag{4.17}$$

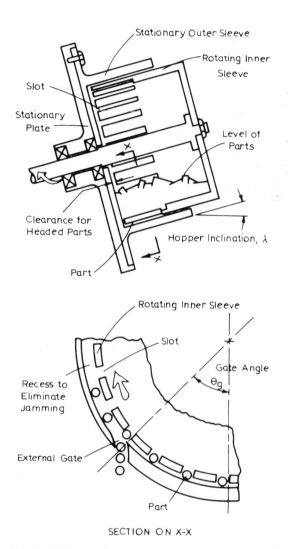

Fig. 4.12 External gate hopper. General data: maximum peripheral velocity, 0.5 m/ s; gate angle θ_g, 0.79–1.13 rad (45–65 deg); hopper inclination λ, 0.17–0.26 rad (10–15 deg).

To give the largest values of v, the gap h_g between the cylinder and sleeve should be as large as possible. For values of h_g greater than $D/2$, there is a danger that the parts may become jammed between the corner B in the slot and the inner surface of the sleeve. Thus, taking $h_g = D/2$, Eq. (4.17) becomes, after rearrangement,

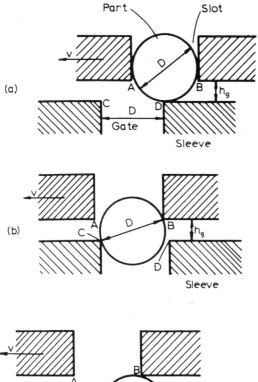

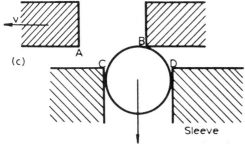

Fig. 4.13 Various stages in the motion of a part passing through the gate of an external gate hopper.

$$v = 0.802(Dg)^{1/2} \tag{4.18}$$

If a_s is now taken to be the centerline distance between adjacent slots of the cylinder, the maximum feed rate F_{max} from the feeder is

$$F_{max} = \frac{v}{a_s} = 0.802(Dg)^{1/2} \tag{4.19}$$

In general, not all of the slots will contain parts and, if E is taken to be the efficiency of the feeder, the actual feed rate is given by

$$F = 0.802(Dg)^{1/2}\frac{E}{a_s} \tag{4.20}$$

The type of feeder analyzed above is generally used for feeding rivets, where the slots in the inner cylinder are open-ended to allow for the rivet heads. If the diameter of the rivet head is D_h, the minimum theoretical distance between the centers of the slots is D_h, and Eq. (4.20) becomes

$$F = \frac{0.802(Dg)^{1/2}E}{D_h} = 0.802E \left[\frac{D}{D_h}\right]\left[\frac{g}{D}\right]^{1/2} \tag{4.21}$$

and, for plain cylindrical parts,

$$F = 0.802E \left[\frac{g}{D}\right]^{1/2} \tag{4.22}$$

where the partition between parts is very small.

This analysis has considered the maximum feed rate from an external gate feeder. In practical designs, the actual feed rate will be less than this because of mechanical limitations, but the result indicates that certain trends might be expected from this type of feeder, and these are now summarized:

1. The maximum unrestricted feed rate is inversely proportional to the square root of the diameter of a cylindrical part.

2. When feeding rivets with the same head diameter, the maximum feed rate is proportional to the square root of the shank diameter; when feeding rivets where the ratio of shank diameter to head diameter is fixed, the maximum feed rate is inversely proportional to the square root of the shank diameter.

3. If a high feed rate is required, the slots in the inner cylinder of the feeder should be as close as possible.

4.4.2 Load Sensitivity and Efficiency

The unrestricted feed rate for a given design of feeder depends on the efficiency E. This may be affected by the load in the hopper, the angle of inclination of the feeder axis, and the position of the external gate. Tests have shown that the most significant of these variables is the angle of inclination λ of the feeder axis. The results presented in Fig. 4.14 indicate that λ should be as low as possible for maximum efficiency.

However, it should be realized that a practical limitation exists because, as λ is reduced, the capacity of the hopper is also reduced, and a compromise must therefore be reached in any given design. A typical figure for λ is between 10 and 15 deg. The results in Fig. 4.14 also show that the efficiency of the feeder increases rapidly as the hopper empties. Figure 4.15 shows the

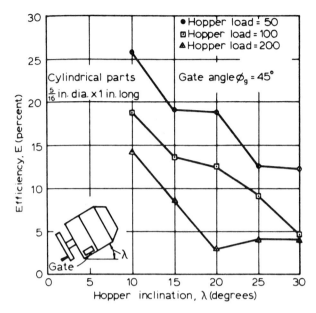

Fig. 4.14 Effect of hopper inclination on the efficiency of an external gate hopper.

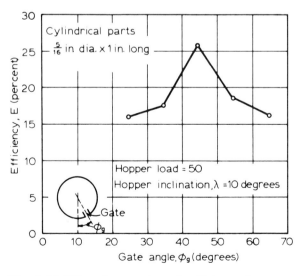

Fig. 4.15 Effect of gate angle on the efficiency of an external gate hopper.

effect of the angular position ϕ_g of the external gate on the efficiency. It is clear that, for values of ϕ_g greater than 90 deg, the efficiency would become zero and, for very small values of ϕ_g, the chances of parts falling into the slots are reduced. The results show that an optimum exists when ϕ_g is approximately 45 deg, and Fig. 4.14 shows that, with this optimum value and with the lowest practical value of λ of 10 deg, the minimum efficiency is 0.14. This represents [from Eq. (4.20) when $a_s = 2D$] a maximum possible feed rate of approximately 2 parts/sec with cylindrical parts of 12.3-mm (5/16-in.) diameter.

Figure 4.16 shows the application of the external gate hopper to the feeding of rivets. In the figure, the feed rate per slot FR_s is plotted against the peripheral velocity of the sleeve v.

4.5 ROTARY-DISK FEEDER

A typical rotary-disk feeder is illustrated in Fig. 4.17. This feeder consists of a disk with a number of slots machined radially in its face and mounted at a steep angle to the horizontal, so that it forms the base of a stationary hopper.

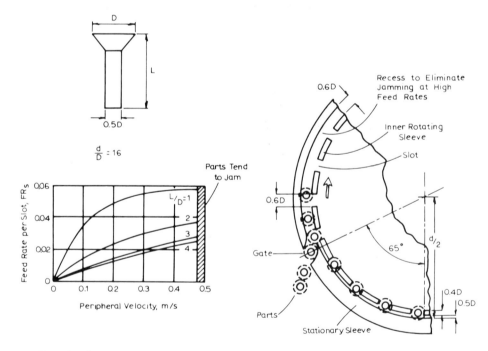

Fig. 4.16 Performance of an external gate hopper when feeding headed parts (adapted from Ref. 1).

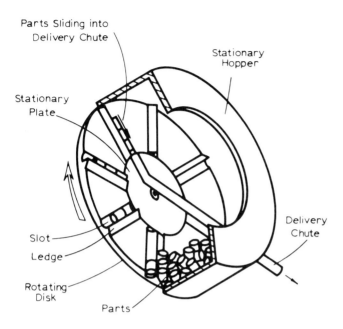

Fig. 4.17 Rotary-disk feeder.

As the disk rotates, the parts in the hopper are disturbed by the ledges next to the slots. Some parts are caught in the slots and carried around until each slot, in turn, reaches the highest position, at which point it becomes aligned with a delivery chute down which the parts slide. A stationary circular plate at the center of the disk prevents the parts from sliding out of the slots until they are aligned with the chute.

In some designs of rotary-disk feeders, the length of the slots allows more than one part per slot to be selected during each revolution of the disk. This design will be analyzed first, and it will be assumed that, to give the greatest efficiency, the disk is indexed with sufficient dwell to allow all the parts selected in each slot to slide down the chute.

4.5.1 Indexing Rotary-Disk Feeder

If a Geneva mechanism is employed to index a rotary-disk feeder, the time for index will be approximately equal to the dwell period. For the design illustrated in Fig. 4.17, the time t_s required for all parts in one slot to slide into the delivery chute is given by Eq. (4.13), which is

$$t_s^2 = \frac{2\ell}{g(\sin\theta - \mu_d\cos\theta)} \qquad (4.23)$$

where ℓ is the length of the slot, θ the inclination of the delivery chute, and μ_d the coefficient of dynamic friction between the part and the chute.

With a Geneva drive, the total period of an indexing cycle t_i is therefore given by

$$t_i = 2t_s = \left[\frac{8\ell}{g(\sin\theta - \mu_d\cos\theta)} \right]^{1/2} \tag{4.24}$$

If L is the length of a part, the maximum number that may be selected in a slot is ℓ/L. In practice, however, the average number selected will be less than this. If E is taken to be the efficiency of the feeder, the feed rate F will be given by

$$F = \frac{E\ell}{Lt_i} = E\left[\frac{\ell g(\sin\theta - \mu_d\cos\theta)}{8L^2} \right] \tag{4.25}$$

It can be seen from Eq. (4.25) that, if E is assumed to be constant:

1. The feed rate is independent of the number of slots in the disk.
2. For a given feeder, the feed rate is inversely proportional to the length of the part.
3. For maximum feed rate with a given part, μ_d should be as low as possible, and both the delivery chute angle θ and the slot length ℓ should be as large as possible.

It is clear, however, that, with the design under consideration, the feed rate will be reduced as the hopper gradually empties until the hopper is almost empty and no more than one part may be selected in each slot.

4.5.2 Rotary-Disk Feeder with Continuous Drive

A rotary-disk feeder with continuous drive would be most suitable for feeding disk-shaped parts. In this case, the analysis for the maximum feed rate would be similar to that for an external gate hopper because the situation in which the part slides from the slot into the delivery chute is similar to that shown in Fig. 4.13. With this device, the slot length is equal to the diameter of the part D, and only one part can be selected in each slot. If the rotational speed of the disk is too high, the parts will pass over the mouth of the delivery chute, and feeding will not occur. If the effect of friction is considered negligible because of the large angle of inclination of the disk, the feed rate at maximum rotational speed will be given by Eq. (4.22), which is

$$F = 0.802E\left(\frac{g}{D} \right)^{1/2} \tag{4.26}$$

where E is the efficiency of the feeder (that is, the average number of parts selected per cycle divided by the number of slots).

The foregoing analyses have considered the theoretical maximum feed rate from a rotary-disk feeder, both with indexing drive and with continuous drive. The results indicate that the following trends may be expected from this type of feeder:

1. For an indexing rotary-disk feeder with long slots, the maximum feed rate is inversely proportional to the length of the part and proportional to the square root of the slot length. For high feed rates, the slope of the delivery chute should be as large as possible, and the coefficient of friction between the part and the chute should be as low as possible.

2. For a feeder with continuous drive, the maximum feed rate for disk-shaped parts is inversely proportional to the square root of the diameter of the part.

4.5.3 Load Sensitivity and Efficiency

Tests were conducted on an indexing rotary-disk feeder with eight slots, each able to carry two cylindrical parts 25 mm (1 in.) in length and 8 mm (5/16 in.) in diameter. The results are presented in Fig. 4.18, which shows that, as expected, the efficiency decreases as the hopper empties. This is because, for small loads, the mass of parts only partly covers the slots and, at best, only one part per slot can be selected during each cycle. The figure shows that both the efficiency and the load-sensitivity characteristics are improved as the angle of inclination θ of the disk is reduced. Unfortunately, this also reduces the inclination θ of the delivery chute and increases the time taken for the parts to slide out of the slots. Clearly, in any given design, a compromise is necessary. In practice, typical values for the angles of inclination are 55–69 deg for the disk and 35 deg for the delivery chute. Figure 4.19 shows an application of the rotary-disk feeder to the feeding of disks.

4.6 CENTRIFUGAL HOPPER FEEDER

The centrifugal hopper feeder shown in Fig. 4.20 is particularly suitable for feeding plain cylindrical parts. In this device, the parts are placed in a shallow cylindrical hopper whose base rotates at constant speed. A delivery chute is arranged tangentially to the stationary wall of the hopper, and parts adjacent to this wall that have become correctly oriented because of the general circulation pass into the delivery chute. No orienting devices are provided in the hopper and, therefore, parts must be taken off in the attitude that they naturally adopt in the hopper, as indicated in the figure.

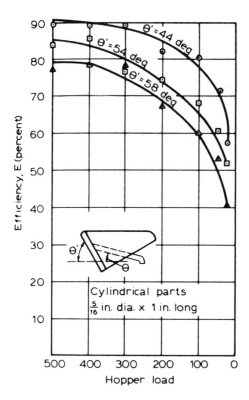

Fig. 4.18 Load sensitivity of a rotary-disk feeder.

4.6.1 Feed Rate

If a part is moving with constant velocity v around the inside wall of a centrifugal hopper, the radial reaction at the hopper wall is equal to the centrifugal force $2m_p v^2/d$, where m_p is the mass of the part and d the diameter of the hopper. The frictional force F_w at the hopper wall tends to resist the motion of the part and is given by

$$F_w = \frac{2\mu_w m_p v^2}{d} \tag{4.27}$$

where μ_w is the coefficient of friction between the part and the hopper wall. When the peripheral velocity of the spinning disk is greater than v, the disk slips under the part, and the frictional force F_b between the part and the spinning disk is given by

$$F_b = \mu_b m_p g \tag{4.28}$$

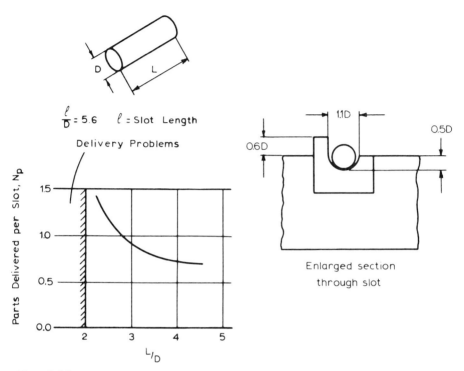

Fig. 4.19 Performance of a rotary-disk feeder when feeding disks (adapted from Ref. 1).

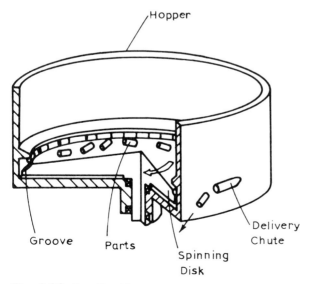

Fig. 4.20 Centrifugal hopper.

where μ_b is the coefficient of friction between the part and the spinning disk. Since, under this condition, $F_b = F_w$, setting Eq. (4.27) equal to Eq. (4.28) gives

$$v = \left[\frac{g\mu_b d}{2\mu_w} \right]^{1/2} \tag{4.29}$$

and the maximum feed rate F_{max} of parts of length L is given by

$$F_{max} = \frac{v}{L} = \frac{(g\mu_b d/2\mu_w)^{1/2}}{L} \tag{4.30}$$

and the actual feed rate F may be expressed as

$$F = E \left[\frac{(g\mu_b d/2\mu_w)^{1/2}}{L} \right] \tag{4.31}$$

where E is the feeder efficiency.

Equation (4.31) shows that the unrestricted feed rate from a centrifugal hopper is proportional to the square root of the hopper diameter and inversely proportional to the length of the parts.

Using Eq. (4.29), the maximum rotational frequency n_{max} of the spinning disk, above which no increase in feed rate occurs, is

$$n_{max} = \frac{v}{\pi d} = \frac{[(g/2d)(\mu_b/\mu_w)]^{1/2}}{\pi} \tag{4.32}$$

This equation is plotted in Fig. 4.21 and can be used to choose the maximum rotational frequency of the hopper.

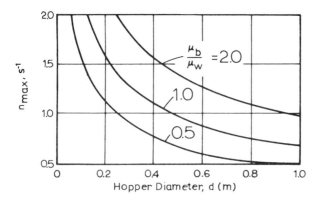

Fig. 4.21 Maximum rotational frequency for a centrifugal hopper, where n_{max} is the maximum rotational frequency of the disk, μ_b the coefficient of friction between the part and the spinning disk, and μ_w the coefficient of friction between the part and the stationary hopper wall.

4.6.2 Efficiency

The overall efficiency E of the hopper can be determined only by experiment. Figure 4.22 plots results of experiments showing the effect of rotational frequency on efficiency when plain cylinders are fed.

4.7 STATIONARY-HOOK HOPPER FEEDER

In the stationary-hook hopper parts feeder (Fig. 4.23) the curved hook is stationary and the concave base of the hopper rotates slowly beneath it. The parts are guided along the edge of the hook toward a ledge at the periphery of the hopper, where they are eventually deflected into the delivery chute by a deflector mounted on the hopper wall. One advantage of this type of feeder is its gentle feeding action, which makes it suitable for feeding delicate parts at low speed.

4.7.1 Design of the Hook

Ideally, the stationary hook should be designed so that a part travels along its leading edge with constant velocity. With reference to Fig. 4.24a, the velocity of the part relative to the hook is designated v_{12}, and the velocity of a point on the moving surface of the base relative to the hook is designated v_{32}. The direction of the velocity v_{13} of the part relative to the base is then obtained from the velocity diagram, Fig. 4.24b.

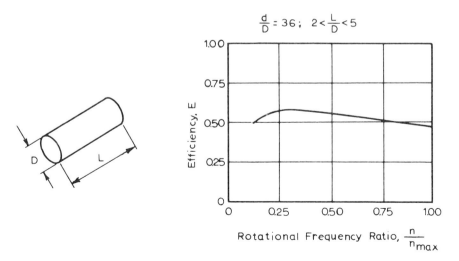

Fig. 4.22 Performances of a centrifugal hopper when feeding cylinders (d = hopper diameter) (adapted from Ref. 1).

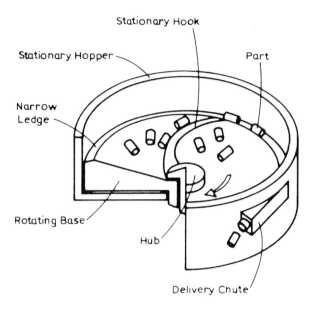

Fig. 4.23 Stationary-hook hopper.

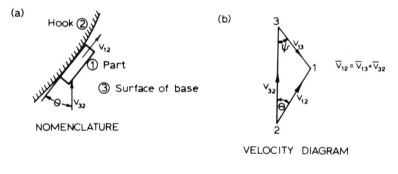

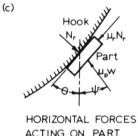

Fig. 4.24 Analysis of stationary-hook hopper.

The frictional force between the part and the moving surface is in the direction opposite to the relative velocity v_{13} and is the force that causes the part to move along the hook. The direction of this force is specified by ψ in Fig. 4.24b. By the law of sines,

$$\frac{v_{12}}{v_{32}} = \frac{\sin \psi}{\sin (\theta + \psi)} \tag{4.33}$$

where θ is the angle between the velocity vector v_{32} and the tangent to the hook at the point under consideration.

Neglecting the inertia force, the external forces acting on the part in the horizontal plane are shown in Fig. 4.24c. When these forces are resolved parallel and normal to the hook tangent, the conditions for equilibrium are

$$\mu_r N_r = \mu_p W \cos(\theta + \psi) \tag{4.34}$$

$$N_r = \mu_p W \sin(\theta + \psi)$$

where μ_r is the coefficient of dynamic friction between the part and the hook, μ_p the coefficient of sliding friction between the part and the moving surface, N_r the normal reaction between the part and the hook, and W the weight of the part. Combining Eqs. (4.34) gives

$$\mu_r = \cot (\theta + \psi) \tag{4.35}$$

Equations (4.33) and (4.35) can be combined to eliminate ψ:

$$\frac{v_{12}}{v_{32}} = \cos \theta - \mu_r \sin \theta \tag{4.36}$$

The velocity of a point on the base relative to the hook is given by

$$v_{32} = 2\pi r n \tag{4.37}$$

where r is the distance from the point under consideration to the center of the hopper and n the rotational frequency of the hopper base. For constant speed of the part along the hook, Eqs. (4.36) and (4.37) can be combined to give

$$r(\cos \theta - \mu_r \sin \theta) = K \tag{4.38}$$

where K is a constant given by $v_{12}/2\pi n$.

An explicit equation defining the shape of the hook is not readily obtainable from Eq. (4.38). However, a numerical method was used [2] to develop the hook shapes, giving constant-speed feeding for various values of the coefficient of sliding friction between the part and the hook; these hook shapes are presented in Fig. 4.25. A computer program constructed the hook shapes, using straight-line segments starting at the hub and working outward. The results in Fig. 4.25 are plotted in a dimensionless form in which the radius to a point on the hook is divided by the hub radius.

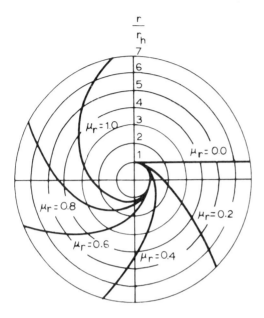

Fig. 4.25 Hook shapes for a stationary hook hopper, where r_h is the radius of the hub and r is the radial location of a particular point on the hook. Maximum peripheral velocity is 0.4 m/s (from Ref. 2).

In using these results to design the shape of a hook for a particular application, it is always preferable to use a larger-than-expected value of the coefficient of sliding friction between the part and the hook. This ensures that parts will not decelerate as they move along the hook and thus tends to avoid jamming in the feeding process.

4.7.2 Feed Rate

It is apparent that the velocity of the parts along the hook will determine the maximum feeder output. For parts of a given length being fed end to end, the maximum output per unit of time is determined by dividing the velocity at the hub by the individual part length. Figure 4.26 shows how the maximum feed rate from the feeder varies with the length of parts for various rotational frequencies.

For particular applications, the actual feed rate will be lower than that given in Fig. 4.26. For a particular design of feeder, this feed rate can be expressed by

$$F = \frac{E\pi dn}{L} \qquad (4.39)$$

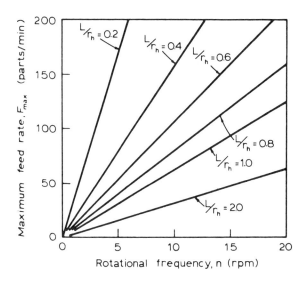

Fig. 4.26 Maximum feed rate for a stationary-hook hopper, where $F = (r_h L)(2\pi n)$, with L the length of the part and r_h the feeder hub radius.

where d is the hopper diameter, n the rotational frequency of the disk, L the part length, and E the efficiency of the feeder. Figure 4.27 illustrates an application of the stationary-hook hopper to the feeding of cylinders, where the effect of part proportions on the feeder efficiency is presented.

4.8 BLADED-WHEEL HOPPER FEEDER

In the bladed-wheel hopper feeder (Fig. 4.28), the tips of the blades of a vertical multibladed wheel run slightly above a groove in the bottom of the hopper. The groove has dimensions such that the parts in the hopper may be accepted by the groove in one particular orientation only. Rotation of the wheel agitates the parts in the hopper and causes parts arriving at the delivery point in the wrong attitude to be pushed back into the mass of parts.

The angle of inclination of the track is generally about 45 deg, and the maximum linear velocity of the blade tip is around 0.4 m/s. Figure 4.29 shows an application of the bladed-wheel hopper to the feeding of cylinders.

4.9 TUMBLING-BARREL HOPPER FEEDER

In the tumbling-barrel hopper feeder (Fig. 4.30), the cylindrical container, which has internal radial fins, rotates about a horizontal vibratory feed track.

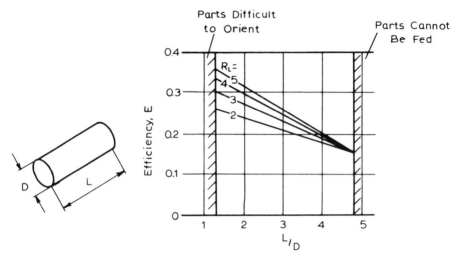

Fig. 4.27 Performance of a stationary-hook hopper when feeding cylinders, with load ratio $R_L = LN/\pi D$, $d/D = 36$ and N is the number of parts in the hopper. (adapted from Ref. 1).

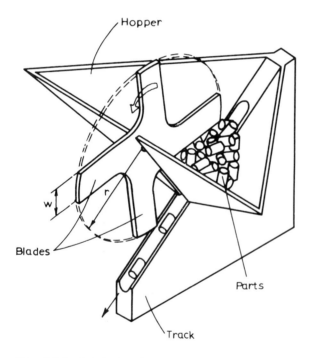

Fig. 4.28 Bladed-wheel hopper.

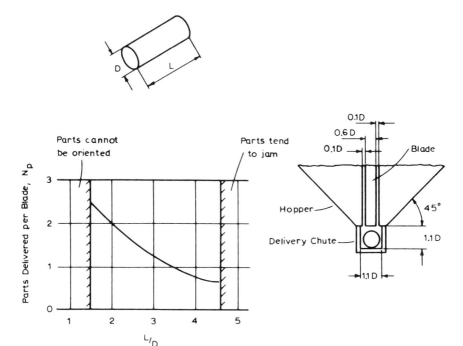

Fig. 4.29 Performance of bladed-wheel hopper when feeding cylinders (adapted from Ref. 1).

Parts placed in bulk in the hopper are carried upward by the fins until, at some point, they slide off the fin and cascade onto the vibratory feed track. The feed track is shaped to suit the required orientation of the part being fed and retains and feeds those parts only in this orientation. This feeder is suitable for feeding a wide variety of parts, such as cylinders, U-shaped parts, angled parts, and prisms.

The optimum hopper load is that which fills the barrel to a height of one-fourth the inside diameter. For parts that can stack one on another, it is necessary to provide the outlet with a stationary gate, which outlines the shape of a part on the rail so that parts come out in a single layer only. Rubber or cork may be placed along the inside wall of the barrel to reduce noise and damage to parts.

4.9.1 Feed Rate

For a part whose motion relative to the vanes is one of pure translation (that is, the part does not roll), the rotational frequency that maximizes the number of parts that land on a rail passing through the barrel center can be determined

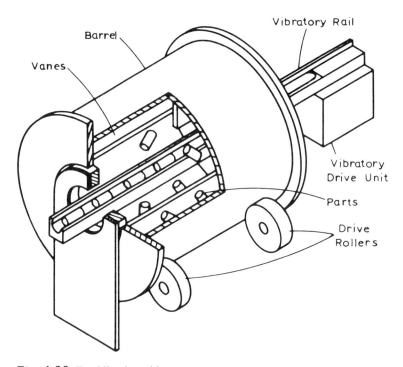

Fig. 4.30 Tumbling-barrel hopper.

from Fig. 4.31a. The result may not be valid for parts whose motion relative to the vanes contains rotational components.

Figures 4.31b–e can be used to estimate the values of the barrel design parameters that achieve a given feed rate. To use the graphs, it is first necessary to determine, by observation, the average number of parts that land on an empty rail per vane P.

Figures 4.32 and 4.33 show applications of the tumbling-barrel hopper to the feeding of cylinders and U-shaped parts, respectively.

4.10 ROTARY-CENTERBOARD HOPPER FEEDER

This feeder (Fig. 4.34) consists of a bladed wheel that rotates inside a suitably shaped hopper. The edges of the blades are profiled to collect parts in the desired attitude and lift them clear of the bulk of the parts. Further rotation of the wheel causes the oriented parts to slide off the blade, which is then aligned with the delivery chute. It is usual to drive the wheel intermittently by either a Geneva mechanism or a ratchet-and-pawl mechanism. The design of the indexing mechanism should take into account the dwell time required for a full

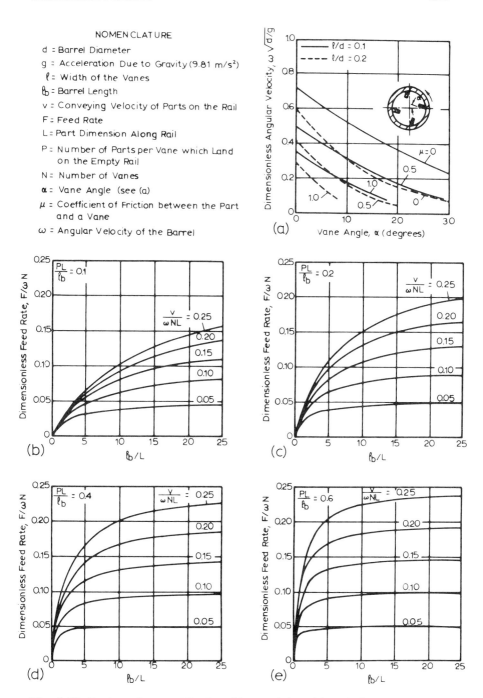

Fig. 4.31 Feed rate for a tumbling-barrel hopper (adapted from Ref. 1).

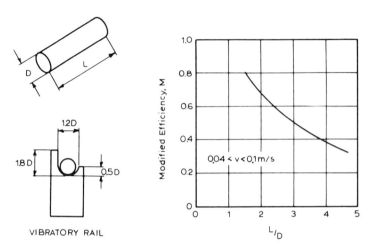

Fig. 4.32 Performance of a tumbling-barrel hopper when feeding cylinders. The feed rate is $Mv/L + 0.7$ parts/sec, where M is the modified efficiency, v the conveying velocity on the vibratory rail (m/s), and L the part length (m). The ratio of barrel length to diameter d is 1.2; $d/D = 21$ (adapted from Ref. 1).

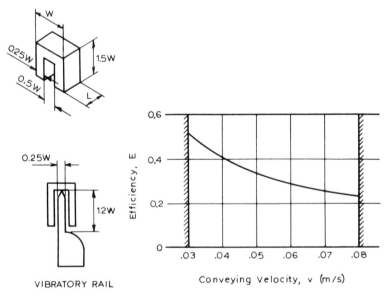

Fig. 4.33 Performance of a tumbling-barrel hopper when feeding U-shaped parts. The feed rate is Ev/L, where E is the efficiency, v the conveying velocity on the vibratory rail, and L the part length. The ratio of barrel length to diameter d is 1.2; $d/W = 13$; and $0.4 < L/W < 1.2$ (adapted from Ref. 1).

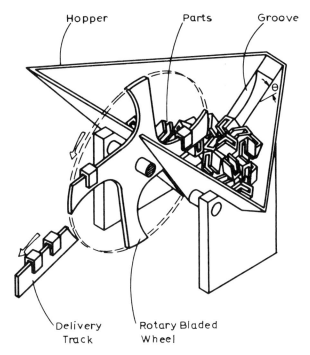

Hopper   Parts   Groove

Delivery   Rotary Bladed
Track   Wheel

Fig. 4.34 Rotary centerboard hopper.

blade to discharge all its parts when aligned with the delivery chute. An analysis similar to that used for the reciprocating-centerboard hopper feeder gives the minimum values for dwell and indexing times and, hence, the maximum feed rate.

The peripheral velocity of the rotary bladed wheel should be no greater than 0.6 m/s. Figure 4.35 shows the application of the feeder to the feeding of U-shaped parts.

4.11 MAGNETIC-DISK FEEDER

This feeder (Fig. 4.36) consists of a container that is closed at one side by a vertical disk. The disk rotates about a horizontal axis, and permanent magnets are inserted in pockets around its periphery. As the disk rotates, parts are lifted by the magnets and stripped off at a convenient point. This feeder can clearly be used only for parts of a ferromagnetic material.

The magnets should have a holding capacity of 10–20 times the weight of one part and a diameter approximately equal to the major dimension of the

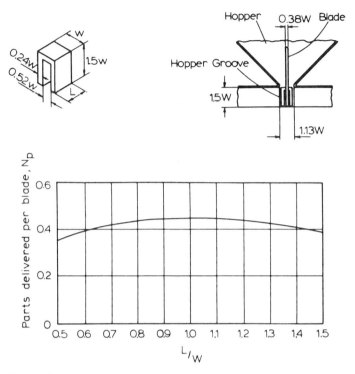

Fig. 4.35 Performance of a rotary centerboard hopper when feeding U-shaped parts (adapted from Ref. 1).

part. The linear velocity of the magnets should be 0.08–0.24 m/s. Figure 4.37 shows an application of the feeder to the feeding of disks of various proportions.

4.12 ELEVATING HOPPER FEEDER

This feeder (Fig. 4.38) has a large hopper with inclined sides. Often, an agitating device is fitted to the base to encourage the parts to slide to the lowest point in the hopper. An endless conveyor belt, fitted with a series of selector ledges, is arranged to elevate parts from the lowest point in the hopper. The ledges are shaped so that they will accept parts only in the desired attitude. The parts slide off the ledges into the delivery chute, which is situated at a convenient point above the hopper. Figure 4.39 shows the application of this feeder to the feeding of cylinders.

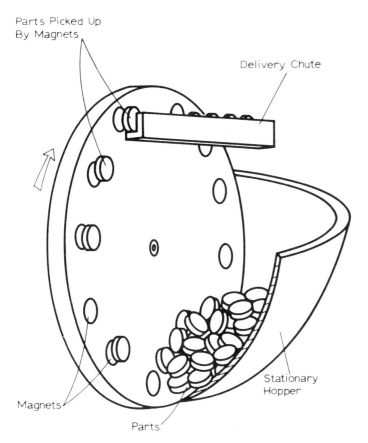

Fig. 4.36 Magnetic-disk feeder.

4.13 MAGNETIC ELEVATING HOPPER FEEDER

The magnetic elevating hopper feeder (Fig. 4.40) is basically the same as the elevating hopper feeder except that, instead of ledges, permanent magnets are fitted to the endless belt. Thus, the feeder is suitable only for handling ferromagnetic materials and cannot easily be used for orientation purposes. With this feeder, it is usual to strip the parts from the magnets at the top of the belt conveyor.

4.14 MAGAZINES

An alternative means of delivering parts to an automatic assembly machine is to use a magazine. With this method, parts are stacked into a container or

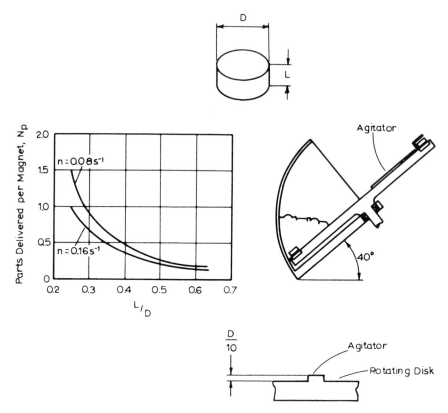

Fig. 4.37 Performance of magnetic-disk feeder when feeding disks (n = rotational frequency) (adapted from Ref. 1).

magazine that constrains the parts in the desired orientation. The magazine is then attached to the workhead of the assembly machine. The magazines may be spring-loaded to facilitate delivery of the parts or, alternatively, the parts may be fed from the magazine under gravity or assisted by compressed air.

Magazines have several advantages over conventional parts feeders and some of these are described below:

1. In some cases, magazines may be designed to accept only those parts that would be accepted by the assembly machine workhead and thus can act as inspection devices. This can give considerable reduction in the downtime on the assembly machine.

2. Magazines can often replace not only the parts feeder but also the feed track.

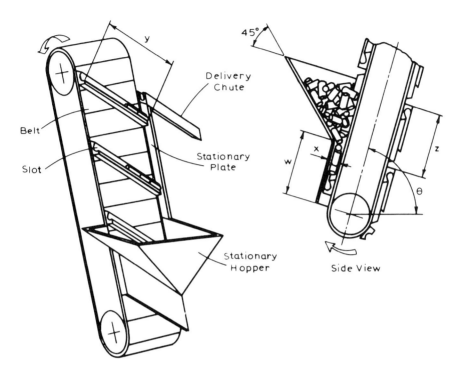

Fig. 4.38 Elevating hopper feeder.

3. Magazines are usually very efficient feeding devices, and assembly machine downtime due to feeder or feedtrack blockages can often be eliminated by their use.

Some benefits may be obtained if the magazines are loaded at the assembly factory. If a number of similar parts are to be used in an assembly, it may be possible to use one hopper feeder to load all the required magazines. Further, if the magazine is designed to accept only good parts, or if some method of inspecting the parts is incorporated into the parts feeder, downtime will occur on the magazine loader and not on the assembly machine.

Some of the disadvantages associated with the use of magazines are as follows:

1. Magazines will generally hold considerably fewer parts than the alternative parts feeder, and magazine changes must therefore be made more frequently than the refilling of the parts feeder.

2. The most suitable place to load the magazines is the point at which the part is manufactured since it is at this point that the part is already oriented.

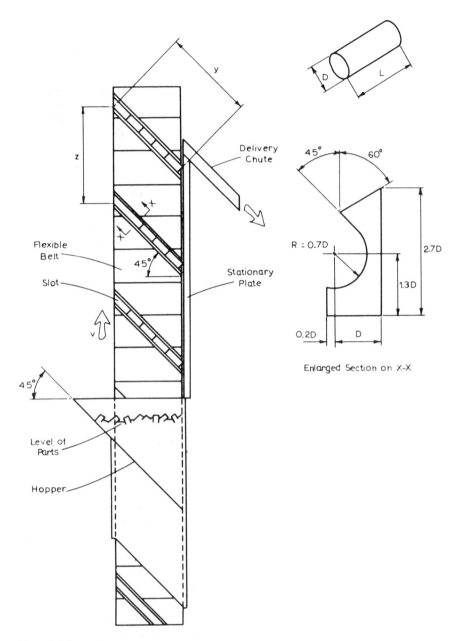

Fig. 4.39 Performance of an elevating hopper feeder when feeding cylinders. The feed rate is 0.6 yv/zL, where v is the velocity of the belt, y the length of one slot, z the distance between two slots, and L the part length; $L/D > 2$; $y/l > 4$ (adapted from Ref. 1).

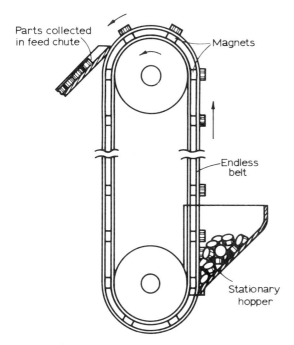

Fig. 4.40 Magnetic elevating hopper feeder.

When manufacture and assembly take place in the same factory, this may not present a serious problem but, if the parts are purchased from another firm, it will be much more difficult to arrange for magazine loading.

3. The cost of magazines can be considerable and prohibitive. The cost and life expectancy should be included in any economic analysis.

In some cases, it is possible to use disposable magazines such as those employed in the packaging of certain drugs. Such a magazine can be fed to the workhead, which would have a suitable mechanism for removing the parts. Another alternative is to blank parts from a strip. In this case, the final operation of separating the parts from the strip can be left to the assembly machine workhead. A further alternative often used for small blanked parts is for the blanking operation to take place on the machine just prior to the point at which the part is required in the assembly.

REFERENCES

1. Boothroyd, G., Poli, C., and Murch, L. E., *Handbook of Feeding and Orienting Techniques for Small Parts*, University of Massachusetts, Amherst, Mass., 1977.
2. Boothroyd, G., and Zinsmeister, G. E., "Design of Stationary Hook Hoppers for Feeding Delicate Parts," *Automation*, May 1969, pp. 64–66.

Feed Tracks, Escapements, Parts-Placement Mechanisms, and Robots

To provide easy access to automatic workheads and the assembly machine, the parts feeders are usually placed some distance away from the workheads. The parts, therefore, have to be transferred and maintained in their orientation between the feeder and the workhead by using a feed track. Most parts feeders do not supply parts at the discrete intervals usually required by an automatic workhead. As a result, the parts feeder must be adjusted to overfeed slightly, and a metering device, usually referred to as an *escapement*, is necessary to ensure that parts arrive at automatic workheads, at the correct intervals. After leaving the escapement, the parts are then placed in the assembly, a process usually carried out by a parts-placing mechanism or a robot.

5.1 GRAVITY-FEED TRACKS

Feed tracks may be classified as either gravity tracks or powered tracks. The majority of tracks are of the gravity type, and these may take many forms. Two typical track arrangements are illustrated in Fig. 5.1; the choice of the design generally depends on the required direction of entry of the part into the workhead. In design, it should be remembered that the track may not always be full, and it is desirable that feeding should still take place under this condition.

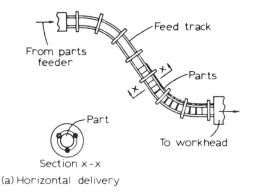

Section x-x

(a) Horizontal delivery

(b) Vertical delivery

Fig. 5.1 Gravity-feed track arrangements.

When the track is partly full and no pushing action is obtained by air jets or vibration, it is clear that the vertical-delivery track design show in Fig. 5.1b will deliver parts from rest at a greater rate than the horizontal-delivery design shown in Fig. 5.1a. The performance of the vertical-delivery track will also be independent of the loading in the track and, if no further parts are fed into it, the track will deliver the last part as quickly as the first. The time of delivery t_p will be given by the time taken for a part to fall a distance equal to its own length.

Thus,

$$t_p = \left[\frac{2L}{g} \right]^{1/2}$$

(5.1)

where L is the length of the part and g is the acceleration resulting from gravity, or 9.81 m/s^2.

5.1.1 Analysis of Horizontal-Delivery Feed Track

In the track design for horizontal delivery (Fig.5.1a), the last few parts cannot be fed and, even if the height of parts in the track is maintained at a satisfactory level, the delivery time will be greater than that given by Eq. (5.1).

Figure 5.2 shows the basic parameters defining the last portion of a horizontal-delivery gravity-feed track. This portion consists of a horizontal section AB of length L_1, preceded by a curved portion BC of constant radius R which, in turn, is preceded by a straight portion inclined at an angle α to the horizontal. It is assumed in the following analysis that a certain fixed number of parts are maintained in the track above the delivery point. If the length of the straight inclined portion of the track containing parts is denoted by L_2, the number of parts N_p is given by

$$N_p = \frac{L_2 + R\alpha + L_1}{L} \tag{5.2}$$

An equation is now derived giving the time t_p to deliver one part of length L. It is assumed in the analysis that the length of each part is small compared to the dimensions of the feed track and that the column of parts can be treated as a continuous, infinitely flexible rod.

When the escapement opens and the restraining force at A is removed, the column of parts starts to slide toward the workhead. In estimating the accelera-

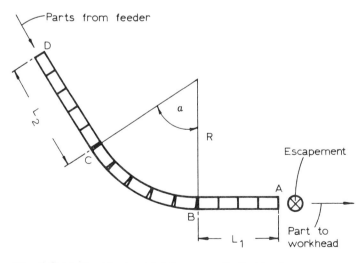

Fig. 5.2 Idealized horizontal-delivery gravity-feed track.

tion a of the column of parts, it is convenient to consider separately the parts in the three sections of the track, AB, BC, and CD, as shown in Fig. 5.3.

If the mass per unit length of the column of parts is denoted by m_1, the weight of section AB is given by $m_1 L_1 g$. The total frictional resistance in this region is given by $\mu_d m_1 L_1 g$ if μ_d is the coefficient of dynamic friction between the parts and the track.

The equation of motion for section AB is given by

$$F_1 = m_1 L_1 (\mu_d g + a) \tag{5.3}$$

where F_1 is the force exerted on the parts in section AB by the remainder of the parts in the feed track and a is the initial acceleration of all the parts.

Similarly, the column of parts in the straight inclined portion CD of the feed track is partly restrained by a force F_2 given by

$$F_2 = m_1 L_2 (g \sin \alpha - \mu_d g \cos \alpha - a) \tag{5.4}$$

To analyze the motion of the parts in the curved section BC of the feed track, it is necessary to consider an element of length $R \, d\theta$ on a portion of the track that is inclined at an angle θ to the horizontal. In this case, a force F resists the motion of the element, and a force $(F + dF)$ tends to accelerate the element. These forces have a small component $F \, d\theta$ that increases the reaction

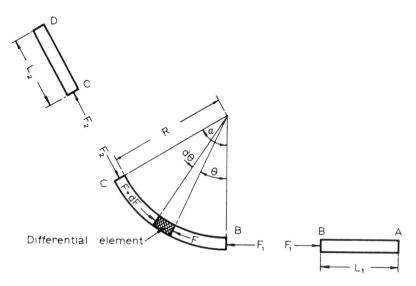

Fig. 5.3 Three separate sections of the idealized horizontal-delivery track.

between the parts and the track and, consequently, increases the frictional resistance.

The forces acting on the element are shown in Fig. 5.4, and the force equilibrium equation in the radial n direction is

$$N_1 = m_1 gR \cos \theta \, d\theta + F \, d\theta \tag{5.5}$$

The equation of motion in the tangential t direction is

$$dF + m_1 gR \sin \theta \, d\theta - \mu_d N_1 = m_1 Ra \, d\theta \tag{5.6}$$

Substituting Eq. (5.5) in Eq. (5.6) and rearranging give the following first-order linear differential equation with constant coefficients:

$$\frac{dF}{d\theta} - \mu_d F = m_1 R(a + \mu_d g \cos \theta - g \sin \theta) \tag{5.7}$$

The general solution to Eq. (5.7) is

$$F = Ae^{\mu_d \theta} + m_1 gR \left[\frac{(1 - \mu_d^2) \cos \theta + 2\mu_d g \sin \theta}{1 + \mu_d^2} - \frac{a}{\mu_d g} \right] \tag{5.8}$$

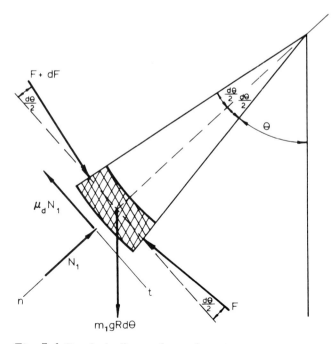

Fig. 5.4 Free-body diagram for an element.

There are two applicable boundary conditions: $\theta = 0$, $F = F_1$, and $\theta = \alpha$, $F = F_2$. The first boundary condition gives

$$A = F_1 - m_1 gR \left[\frac{1 - \mu_d^2}{1 + \mu_d^2} - \frac{a}{\mu_d g} \right] \tag{5.9}$$

where F_1 is given by Eq. (5.3)

Equation (5.8) relates the acceleration of the line of parts along the curved and horizontal sections of the track when a force F is applied to the end of the line located at an angle θ. This applied force at the end is F_2, given by Eq. (5.4). Using this second boundary condition and solving Eq. (5.8) for a/g give

$$\frac{a}{g} = \frac{(L_2/R)(\sin\alpha - \mu_d\cos\alpha) - (L_1/R)\mu_d e^{\mu_d\alpha} + \{[(1-\mu_d^2)(e^{\mu_d\alpha} - \cos\alpha) - 2\mu_d\sin\alpha]/1 + \mu_d^2\}}{[(L_2/R) + (L_1/R)e^{\mu_d\alpha} + [(e^{\mu_d\alpha}-1)/\mu_d]} \tag{5.10}$$

Equation (5.10) shows the relationship between the nondimensional acceleration of the entire line of parts a/g and the design parameters L_2/R, L_1/R, α, and μ_d. Since the part length is small compared to the total length of track, the acceleration is assumed constant during delivery of one part, and thus the time t_p for one part to move into the workhead can be found from the kinematic expression.

$$t_p = \left[\frac{2L}{a} \right]^{1/2} \tag{5.11}$$

where L is the length of the part.

Unfortunately, Eq. (5.10) is cumbersome to use for calculations and, as a consequence, it is not easy to determine quickly the effect of a design change. It is also unrealistic to graph the information contained in Eq. (5.10) since there are five independent nondimensional design parameters. A nomogram would simplify this work, but Eq. (5.10) is not in a form suitable for this approach. However, the following method, developed by Murch and described in Ref. [1], allows a nomogram to be developed that will give results sufficiently accurate for practical purposes. Equation (5.10) can be approximated by

$$\frac{a}{g} = f_1(\mu_d) + f_2\left[\frac{L_1}{R}\right] + f_3(\alpha) + f_4\left[\frac{L_2}{R}\right] \tag{5.12}$$

which is a summation of independent functions of a single variable and is an acceptable form for a nomogram. The functions are

$$f_1 = 0.94(0.5 - \mu_d) \tag{5.13}$$

$$f_2 = -0.428(L_1/R) \tag{5.14}$$

$$f_3 = 0.0084(\alpha - 45) \tag{5.15}$$

with α expressed in degrees, and

$$\frac{L_2}{R} = \frac{0.31e^{9.66(a/g)}}{1 + 0.018e^{-63(a/g)}} \tag{5.16}$$

Although no explicit function f_4 could be found to satisfy Eq. (5.12), a numerical relationship can be developed from Eq. (5.16) for producing a nomogram.

Since Eq. (5.11) is also in a form suitable for nomographic presentation, the combined nomograms for Eqs. (5.10) and (5.11) are shown together in one nomogram in Fig. 5.5. This nomogram relates all the parameters that must be considered for the design of the lower section of a horizontal-delivery gravity-feed track.

5.1.2 Example

Suppose that a workhead is operating at a rate of 1 assembly/sec and has 0.2 sec to receive a 25-mm part; the value of L_1/R is ordinarily chosen as small as possible and, in this situation, equals zero; the value of the dynamic coefficient of friction is 0.5. What are the values of L_2/R and α that will complete this design?

A line is drawn on the nomogram (Fig. 5.5) from a value of part length of 25 mm (1 in.) through a value of time of 0.2 sec and intersects the a/g line at a value of 0.13. A second line is drawn from this value of a/g to a value of zero on the L_1/R scale. This line intersects turning line A. The third line, drawn from turning line A to a value of 0.5 on the μ scale, intersects turning line B. The last line is drawn from this intersection of turning line B through the α and the L_2/R scales. A number of solutions are possible; one such solution gives values of α and L_2/R of 60 deg and 0.27, respectively. Substitution of these design parameters back into Eq. (5.10) yields a value for a/g of 0.12. The subsequent time from Eq. (5.11) is 0.207 sec which, compared to the system design specification of 0.2 sec, is less than a 4% error.

Several comparisons were made to check the accuracy of this nomogram, and the results showed that this nomogram was significantly accurate for design work. It is reasonable to expect the nondimensional acceleration found using this nomogram to be correct within 0.03. The worst results occur with

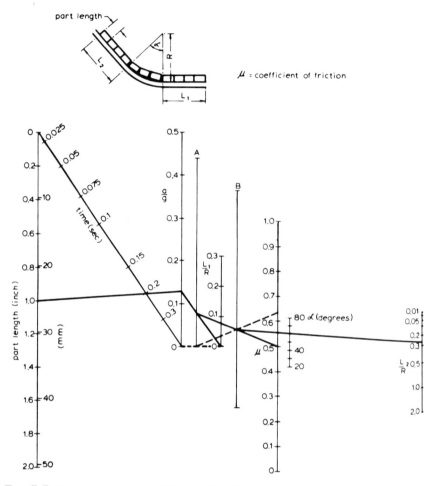

Fig. 5.5 Nomogram solution to Eqs. (5.10) and (5.11).

small angles and large values of L_2/R or large coefficients of friction, but these situations are ordinarily avoided in practice.

It is interesting to note the effect of the value of the coefficient of friction on the value of the nondimensional acceleration a/g. From Eqs. (5.12) and (5.13), it is clear that an error in the value of the coefficient of friction produces a similar error in the value of a/g. Apparently, the source of the greatest error in this work is in the estimation of the value of the coefficient of friction. Some typical values are given in Table 5.1. A simple method for determining the dynamic coefficient of friction has also been developed that has the advantage

Table 5.1 Dynamic Coefficient of Friction of a Variety of Part and Track Material Combinations

Part	Track					
	Nylon	Plexiglas	Brass	Aluminum	Cast iron	Steel
Nylon	0.520	0.536	0.568	0.475	0.375	0.503
Plexiglas	0.502	0.473	0.537	0.503	0.411	0.425
Brass	0.354	0.425	0.370	0.345	0.216	0.250
Aluminum	0.416	0.458	0.437	0.374	0.304	0.327
Cast iron	0.314	0.370	0.368	0.268	0.218	0.252
Steel	0.349	0.419	0.432	0.353	0.273	0.306

of using the actual parts in the experiments. The details are presented in Appendix A.

It has been tacitly assumed that the static friction would not affect this dynamic analysis. If the value of the static coefficient of friction μ_s is significantly large, however, the parts in the queue simply will not move. This occurs if the numerator in Eq. (5.10) is less than or equal to zero when the value of the angle α is greater than arctan μ_s. For this calculation, the value of the static coefficient of friction is substituted for the dynamic coefficient of friction μ_d. Using the design parameters $L_1/R = 0$, $L_2/R = 0.27$, and $\alpha = 60$ deg, the nondimensional acceleration from Eq. (5.10) equals zero when the value of the static coefficient of friction is less than this value.

Fortunately, the nomogram can also be used for this static analysis. A line drawn between $a/g = 0$ and L_1/R intersects turning line A. A line drawn from this intersection through the previous intersection of turning line B crosses the μ line at the critical value of μ_s. Dashed lines are used on Fig. 5.5 to show this procedure for the preceding example.

5.1.3 On-Off Sensors

On automatic assembly machines, it is essential to maintain a supply of correctly oriented parts to the workhead. A workhead that is prepared to carry out the assembly operation but is awaiting the part to be assembled represents downtime for the whole assembly line. Thus, for obvious economic reasons, correctly oriented parts should always be available at each station on an assembly machine. The probability that a part will be available at a workhead will be called the *reliability of feeding* and should not be confused with the problems associated with defective parts, which may jam in the feeding device or workhead and thus affect the reliability of the complete system. The latter subject is discussed in detail in a later chapter.

When automatic feeding devices are used on assembly machines, the mean feed rate is usually set higher than the assembly rate to ensure a continuous supply of parts at the workhead. Under these circumstances, the feed tracks become full, and parts back up to the feeder. For vibratory-bowl feeders, however, this method is not always satisfactory. With these feeders, the parts in the line in the feed track are usually prevented by a pressure brake from interfering with the orienting devices mounted on the bowl-feeder track. If a part filters past the orienting devices and arrives at the pressure break when the delivery chute is full, it is rejected back into the bowl. If this occurs too frequently, excessive wear to both the bowl and the parts can result.

To solve the problems caused by overfeeding and, consequently, ease the problem of wear, a simple control system can be employed that incorporates sensing devices located at two positions on the feed track. The feeder is turned on when the level of parts falls below the sensing device nearest the workhead. The feeder is still set to deliver oriented parts at a higher rate than required, and the parts gradually fill the delivery chute and back up to, and then past, the level of the first sensing device. When the second sensing device, which is located near the feeder, is activated, the bowl is turned off, and the level of parts starts to fall again. When the level of parts falls below the level of the first sensing device, the first device is activated, and the bowl feeder is again set in motion. Since the output of the bowl is a random variable, there may be some time delay before an oriented part leaves the bowl. Thus, it is possible that the entire line of parts between the lower sensor and the critical level found from Eq. (5.10) will be used by the workhead before an oriented part arrives from the feeder. This situation has been studied theoretically [2] to provide the designer of assembly machines with the information necessary to position the sensing devices in the feed track so that the probability of the workhead being starved of parts is kept within acceptable limits.

Theory

Most vibratory feeders are fitted with an orienting system that reorients or rejects those parts that would otherwise be fed incorrectly oriented. It is assumed that, without the orienting system and for a particular setting, a vibratory-bowl feeder would deliver n_f parts during the period of one workhead cycle. Thus, if η is the proportion of parts that pass through the orienting system, then $n_f\eta$ is the average number of parts that pass into the feed track during each workhead cycle. The factor η can be regarded as the efficiency of the orienting system, as discussed in Chapter 3.

The output from the bowl feeder is assumed to have a binominal distribution with a mean of $n_f\eta$ and a variance of $n_f\eta(1 - \eta)$ for each workhead cycle. For large values of n_f, a binomial distribution can be closely approximated by a normal distribution of the same mean and variance. The approximation improves as n_f increases but is always quite good if η is neither close to zero nor close to unity. Thus, the output from the feeder is approximately normal $[n_f\eta, \; n_f\eta(1 - \eta)]$ for every time span of one workhead cycle. The value of $n_f\eta$ must be greater than unity if the average value of the queue length is to increase. In ordinary steady-state queuing theory, $n_f\eta$ is termed the *traffic density* and is less than unity. For this reason, the results of steady-state queuing theory cannot be applied to this problem.

Referring to Fig. 5.6, N_b is the number of parts that can be held in the feed track between the lowest acceptable line level and the first sensing device. Thus, if t_w is the cycle time of the workhead, the distribution of parts that build up the line in $N_b t_w$ sec is approximately normal $[N_b(n_f\eta - 1), \; N_b n_f\eta(1 - \eta)]$, where N_b is subtracted from the mean to allow for the N_b parts that are assembled in $N_b t_w$ sec.

For a particular value of N_b, Fig. 5.7 shows the density function for the net number of parts that build up the line in $N_b t_w$ sec. The value of N_b corresponds to a particular reliability r, which means that the workhead will be starved of parts only $100 \, (1 - r)\%$ of the time following an activation of the on sensor. The area under the density function in Fig. 5.7 equals r. An increase or decrease in N_b changes the mean $[N_b(n_f\eta - 1)]$ more than it changes the standard deviation $[N_b n_f\eta(1 - \eta)]^{1/2}$. Thus, the smaller the value of N_b, the smaller the area r under the density function and the lower the reliability.

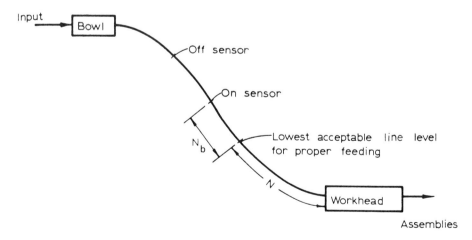

Fig. 5.6 Gravity-feed track with on/off sensors.

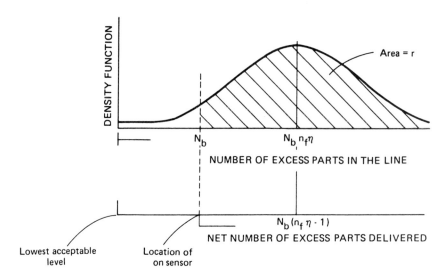

Fig. 5.7 Density functions for delivered parts (from Ref. 2).

Since the area under the curve in Fig. 5.7 is r, the average number of parts to leave the bowl feeder $(N_b n_f \eta)$ is equal to n_s standard deviations $n_s[N_b n_f \eta (1 - \eta)]^{1/2}$, where n_s is determined from a normal distribution table. Thus,

$$N_b = \frac{n_s^2(1 - \eta)}{n_f \eta} \tag{5.17}$$

For example, for a reliability of 0.99997 ($n_s = 4$),

$$N_b = \frac{16(1 - \eta)}{n_f \eta} \tag{5.18}$$

where $n_f \eta$ and η are known parameters for the bowl feeder.

In cases in which these parameters are unknown, it may be convenient to let $N_b = n_s^2$, which is its maximum value since η is always positive and $n_f \eta$ must always be set greater than unity.

It has been stated that, for a binomial distribution to be closely approximated by a normal distribution, n_f should be large. To determine a satisfactory value for n_f for particular values of η and r, a computer simulation routine was developed. Table 5.2 shows some of the results from this simulation for $r = 0.99997$. For relatively large values of η, the corresponding values of $n_f \eta$ may be too large to be of any practical interest. However, when the theoretical

Table 5.2 Simulation Results for Eq. (5.18); $r = 0.99997$

η	n_f	$n_f\eta$
0.2	10	2
0.5	4	2
0.8	20	16

value of N_b was increased by 2, the same simulation yielded the results shown in Table 5.3. Thus, for $n_f\eta \geqslant 1.6(r = 0.9997)$,

$$N_b = \frac{16(1 - \eta)}{n_f\eta} + 2 \tag{5.19}$$

Equation (5.19) can be used to estimate the number of parts between the lower sensor and the level of parts for acceptable operation when the feeder is capable of delivering at a rate 60% greater than the workhead assembly rate, a situation that is not uncommon when on-off sensors are employed. An examination of Eq. (5.19) indicates that the maximum value of N_b is 12. This occurs when η is small and $n_f\eta$ equals 1.6, its minimum. Other conditions that satisfy the constraint on Eq. (5.19) will always produce results for N_b less than or equal to 12. Thus, when little is known about a feeder except that it significantly overfeeds, the lower sensor can be placed in a position in which N_b equals 12 and satisfactory performance is assured.

The number of parts that can be held between sensing devices has no effect on the average running time of the feeder but does affect the average frequency of activations of the sensing devices.

Often, only one sensor is used. Besides activating the feeder, it activates a time delay to simulate the second sensor. Several types of sensors, including fiber optics or pneumatics, can be used on the track, but the basic logic circuitry governing their operation is identical.

Table 5.3 Simulation Results for Eq. (5.19); $r = 0.99997$

η	n_f	$n_f\eta$
0.1	16	1.6
0.2	8	1.6
0.5	3	1.5
0.8	2	1.6

5.1.4 Feed Track Section

A compromise is necessary when designing the feed track section. The clearances between the part and the track must be sufficiently large to allow transfer and yet small enough to keep the part from losing its orientation during transfer. In the curved portions of the track, further allowances have to be made to prevent the part from jamming. Figure 5.8 shows a cylindrical part in a curved tubular track. For the part to negotiate the bend, the minimum track diameter d_t is given by

$$d_t = c + D \qquad (5.20)$$

and by geometry,

$$(R + d_t - c)^2 + \left[\frac{L}{2} \right]^2 = (R + d_t)^2$$

or

$$2c(R + d_t) - c^2 = \left[\frac{L}{2} \right]^2 \qquad (5.21)$$

where R is the inside radius of the curved track and L the length of the part. If c is small compared with $2(R + d_t)$, this expression becomes approximately

$$2c(R + d_t) = \left[\frac{L}{2} \right]^2 \qquad (5.22)$$

Substituting for c from Eq. (5.20) into Eq. (5.22) and rearranging give

$$d_t = 0.5 \left\{ \left[(R + D)^2 + \frac{L^2}{2} \right]^{1/2} - (R - D) \right\} \qquad (5.23)$$

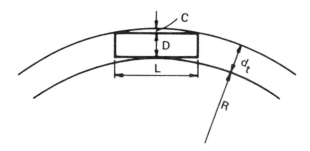

Fig. 5.8 Construction to determine minimum diameter of a curved feed track.

If the parts are sufficiently bent or bowed, it may be difficult to design a curved track that will not allow overlapping of parts and consequent jamming. This is illustrated in Fig. 5.9.

Figure 5.10 illustrates typical track sections used for transferring cylinders, flat plates, and headed parts. An important point to be remembered when designing a feed track is that the effective coefficient of friction between the parts and the track may be higher than the actual coefficient of friction between the two materials. Figure 5.11 gives some examples of the effect of track cross section on the effective coefficient of friction. The increase of 100% in friction given by the example in Fig. 5.11d could have very serious

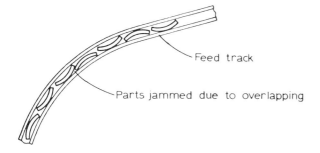

Feed track

Parts jammed due to overlapping

Fig. 5.9 Blockage in a curved feed track.

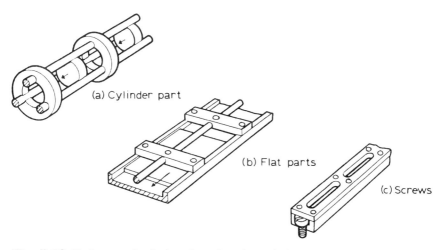

(a) Cylinder part

(b) Flat parts

(c) Screws

Fig. 5.10 Various gravity-feed track sections for typical parts.

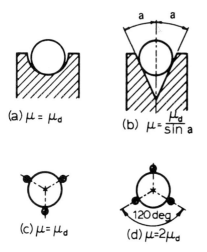

Fig. 5.11 Relation between effective coefficient of friction μ and actual coefficient of friction μ_d for various track designs.

consequences in a gravity-feed system. It is also important to design these tracks with removable covers or access holes for the quick removal of jammed parts.

5.1.5 Design of Gravity-Feed Tracks for Headed Parts

Of the many parts that can be fed in a gravity-feed track, perhaps the most common are headed parts such as screws and rivets, which are often fed in the manner shown in Fig. 5.12. Clearly, if the track inclination is too small or if the clearance above the head is too small, the parts will not slide down the track. It is not always understood, however, that the parts may not feed satisfactorily if the track has too steep an inclination or if too large a clearance is provided between the head of the part and the track. Also, under certain circumstances, a part may not feed satisfactorily, whatever the inclination or clearance. This analysis of the design of gravity-feed tracks for headed parts provides the designer of assembly machines and feeding devices with the information necessary to avoid situations in which difficulty in feeding will occur.

Analysis

Figure 5.13 shows a typical headed part (a cap screw with a hexagon socket) in a feed track. It is clear that, as the track inclination θ is gradually increased and provided that the corner B of the screw head has not contacted the lower

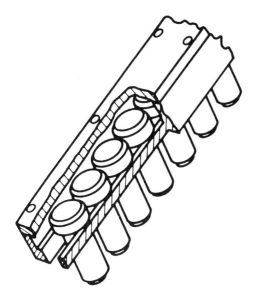

Fig. 5.12 Headed parts in a gravity-feed track.

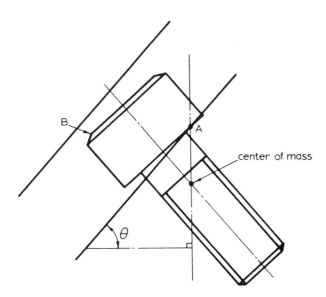

Fig. 5.13 Position of a headed part that does not touch the track cover.

surface of the cover of the track, the screw will slide when $\theta >$ arctan μ_1, where μ_1 is a function of the coefficient of static friction between the screw and the track. On further increases in the track inclination, the condition shown in Fig. 5.14 will eventually arise when the corner B of the screw head has just made contact with the lower surface of the cover of the track. Immediately prior to this condition, the center of mass of the screw lies directly below AA, a line joining the points of contact between the screw head and the track. From Fig. 5.14, it can be seen that z, the distance from the line AA to the axis of the screw, is given by

$$z^2 = \frac{d^2 - s^2}{4} \tag{5.24}$$

where s is the width of the slot and the diameter of the shank and d is the diameter of the screw head.

Also, from the triangle ACE (Fig. 5.14),

$$\frac{z}{\ell} = \tan(\theta_T - \alpha) \tag{5.25}$$

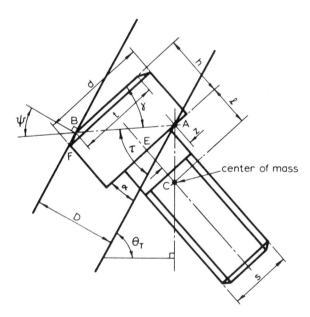

Fig. 5.14 Position of a headed part when corner B just contacts the lower surface of the track cover.

Where ℓ is the distance from the center of mass of the screw to a plane containing the underside of the head, α is the angle between the screw axis and a line normal to the track, and θ_T, which will be called the *tilt angle*, is the track angle at which the top of the screw head just contacts the lower surface of the top cover of the track. From Fig. 5.14,

$$\alpha = \tau - \gamma \tag{5.26}$$

where

$$\tau = \arcsin \frac{D}{[h^2 + (t/2 + z)^2]^{1/2}} \tag{5.27}$$

and

$$\gamma = \arctan \left(\frac{h}{t/2 + z} \right) \tag{5.28}$$

where D is the depth of the track, h the depth of the screw head, and t the diameter of the top of the screw head. Thus, combining Eqs. (5.26–5.28) gives

$$\theta_T = \arctan \left(\frac{z}{\ell} \right) - \arctan \left(\frac{h}{t/2 + z} \right) + \arcsin \left(\frac{D}{[h^2 + (t/2 + z)^2]^{1/2}} \right) \tag{5.29}$$

Sliding will always occur if $\theta_T > \theta > \arctan \mu_1$ since, under these circumstances, the track angle is greater than the angle of friction and there is no contact between the top of the screw head and the track cover. However, sliding may still occur if θ is larger than θ_T. This situation is shown in Fig. 5.15, where it can be seen that a frictional force occurs between the screw head and the track cover as well as on the lower portion of the track.

For sliding to occur under these conditions,

$$mg \sin \theta > F_1 + F_2 \tag{5.30}$$

where

$$F_1 = \mu_1 N_1$$

and $$\tag{5.31}$$

$$F_2 = \mu_2 N_2$$

where μ_1 and μ_2 are the effective values of the coefficient of static friction between the part and the track at A and B, respectively (Fig. 5.15). Resolving forces normal to the track gives

$$N_1 = N_2 + mg \cos \theta \tag{5.32}$$

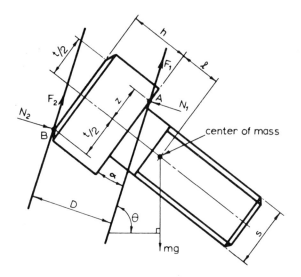

Fig. 5.15 Forces acting on a cap screw in a feed track.

and taking moments about A gives

$$mg[\ell \sin(\theta - \alpha) - z \cos(\theta - \alpha)] \tag{5.33}$$

$$= N_2 \left[\left(\frac{t}{2} + z \right) \cos \alpha - h \sin \alpha \right] - F_2 D$$

Substituting Eqs. (5.31–5.33) in Eq. (5.30) and rearranging give

$$\sin \theta > \mu_1 \cos \theta + (\mu_1 + \mu_2) \frac{\ell \sin(\theta - \alpha) - z \cos(\theta - \alpha)}{[(t/2 + z) \cos \alpha - h \sin \alpha - \mu_2 D]} \tag{5.34}$$

where μ_1 is the effective coefficient of friction between the screw head and the bottom portion of the track, and μ_2 is the effective coefficient of friction between the screw head and the track cover.

Under the conditions shown in Figs. 5.13–5.15, a portion of the screw head lies below the lower contact surface of the track, and the effective coefficient of static friction between the screw and the track is greater than the actual coefficient of static friction μ_s for the two materials. It can be shown that, for this situation, the relation between μ_s and μ_1 is given by

$$\mu_1 = \mu_s \left[\left(d \sin \frac{\alpha}{2z} \right)^2 + \cos^2 \alpha \right]^{1/2} \tag{5.35}$$

In this particular case, the difference between μ_s and μ_1 is small but, in some cases discussed later in this section, this effect is of importance. Since there is point contact between the track cover and the screw head, $\mu_2 = \mu_s$.

Thus, at least two conditions govern the motion of the screw in the track. First, if the track angle is greater than the angle of friction and less than or equal to the tilt angle θ_T, the screw slides. Under these circumstances, the greater the tilt angle θ_T, the larger the range of values of μ_s for which the screw slides and, therefore, as can be seen from Eq. (5.29), the value of D, the track depth, should be as large as possible.

Second, the maximum track angle is restricted to the value given by Eq. (5.34) and, in this case, again, the greater the depth of the track, the greater this maximum value θ. It is of interest now to determine the critical value of track depth below which a screw cannot jam in the track since this will allow a complete definition of the range of track angles and track depths for which a part will feed satisfactorily.

If the track depth D is gradually increased, a special case of the condition shown in Fig. 5.15 will eventually arise. This occurs when the angle between a line joining A and B and a line normal to the track becomes equal to the angle of friction between the screw and the track. For larger values of D, the part will not normally make contact with the upper portion of the track. Under these circumstances, however, the screw will jam in the track if it makes contact across A and B. Thus, the situation arises in which theoretically, the screw should slide but, if a small perturbation rotates it sufficiently to contact the cover, the screw will lock in the track. This possibility should clearly be avoided in practice. From the geometry of Fig. 5.15, the maximum value of D is thus given by

$$D_{\max} = [h^2 + \left(\frac{t}{2} + z\right)^2 \cos \beta_2]^{1/2} \tag{5.36}$$

where $\tan \beta_2 = \mu_2 = \mu_s$.

Figure 5.16 shows a graph of θ plotted versus D/h for a particular screw where $d = 1.712s$, $h = s$, $\ell = 1.5s$, and $t = 1.42s$. These screw dimensions were taken from *Machinery's Handbook* and are the American standard for a size 8 hexagon socket-type cap screw.

Although, in practice, the width of the track s would be larger than the diameter of the screw shank, it should be as small as possible and, in all the examples dealt with in this section, it is considered to be equal to the shank diameter.

The results in the figure show the ranges of values θ and D/h for which the screw will slide down the track without the possibility of jamming for various values of μ_s. It can be seen that, as the coefficient of friction is increased, the

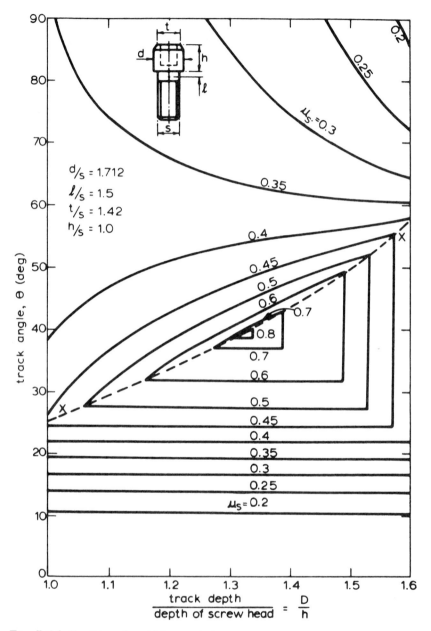

Fig. 5.16 Conditions for which a particular screw will slide in a gravity-feed track (from Ref. 3).

ranges of values of θ and D for feeding to occur decrease. It should also be noted that the line XX, representing the tilt angle θ_T, passes through the points that give (1) the minimum track angle and minimum track depth and (2) the maximum track angle and maximum track depth.

It is now suggested that a reasonable criterion for the best track inclination and track depth would be one in which these parameters are such that feeding would occur for the widest range of values of μ_s. This condition occurs when the points defined under (1) and (2) in the preceding paragraph become identical. Referring to Fig. 5.16, this condition occurs when $\theta_{max} = \theta_{min}$ and $D_{max} = D_{min}$, that is, when $\theta = 39$ deg and $D/h = 1.32$, and, for this situation, it can be seen that the screw will always slide if $\mu_s < 0.81$.

Now θ_{min} occurs when $\theta_T = \beta_1$. Substituting in Eq. (5.25) gives

$$\tan(\beta_1 - \alpha) = \frac{z}{\ell}$$

or

$$\beta_1 - \alpha = \arctan\left(\frac{z}{\ell}\right) \qquad (5.37)$$

When θ equals θ_{max}, it also equals θ_T; hence, from Fig. 5.14, the value of θ_{max} occurs when ψ equals β_2, which gives

$$\tan(\alpha + \beta_2) = \frac{z + t/2}{h}$$

or

$$\alpha + \beta_2 = \arctan\left(\frac{z + t/2}{h}\right) \qquad (5.38)$$

The relationships among β_1, β_2, and β are given by

$$\tan\beta_1 = \tan\left[\left(\frac{d\sin\alpha}{2z}\right)^2 + \cos^2\alpha\right] \qquad (5.39)$$

and

$$\beta_2 = \beta \qquad (5.40)$$

Solving Eqs. (5.37–5.40) simultaneously gives the maximum angle of friction β for which the screw will slide without the possibility of jamming. Substitution of this value in Eqs. (5.39) and (5.36) gives the corresponding optimum values of the track inclination $\theta(= \beta_1)$ and the track depth D.

The equations necessary to determine the maximum coefficient of friction $\mu_{max}(= \tan \beta)$, the track inclination, and the track depth for the proposed optimum conditions for the following four common types of screws have been developed: [3]

1. A cap screw of hexagon socket type
2. A flat-head cap or machine screw in a V track
3. A flat-head cap or machine screw in a plain track
4. A button-head cap or round-head machine screw

In all cases, it has been assumed that the width of the slot in the track is equal to the diameter of the screw shank.

Results

Graphs of μ_{max}, D/h, and θ versus ℓ/s for the four types of screws studied are shown in Figs. 5.17–5.20. Each figure shows these values for the two extreme geometries of the particular type of screw. For example, in the case of a cap-head screw of the hexagon socket type, regardless of its size, all values of d/s and t/s fall within the ranges 1.33–1.712 and 1.041–1.42, respectively.

Clearly, the value of (D/h) cannot be less than unity and, thus, for some types of screws, it can be seen from the figures that there is a minimum value of ℓ/s for which the analysis is valid. If ℓ/s is less than this minimum value, the parts can be fed in a track with a small clearance, and the only restriction on the track angle is that it must be greater than the effective angle of friction.

Although, in practice, there will often be no choice as to the type of screw to be used, it is of interest to note the degree of difficulty in feeding equivalent sizes of the various screws analyzed. For a given length of screw, the most difficult to feed are the larger sizes of cap-head screws of the hexagon socket type and, in this case, if μ_s is 0.5, it would be very difficult to feed the screw if its length were greater than five times its diameter. The easiest screws to feed are the flat-head cap or machine screws and, of the two alternative track designs examined, the V track has better feeding characterisitics.

One general point is that, in many cases, the optimum track depth is very much larger than would normally be used in practice and, when the track depth is only slightly larger than the depth of the head, difficulty is often encountered in feeding these parts.

Procedure for Use of Figs. 5.17-5.20

The first step in using the data provided in the figures is to determine the value of ℓ/s for the screw under consideration. The simplest way is to balance the screw on a knife edge and measure the distance ℓ from the center of mass to the underside or top of the screw head, whichever is appropriate. This value is then divided by the shank diameter to give the ratio ℓ/s.

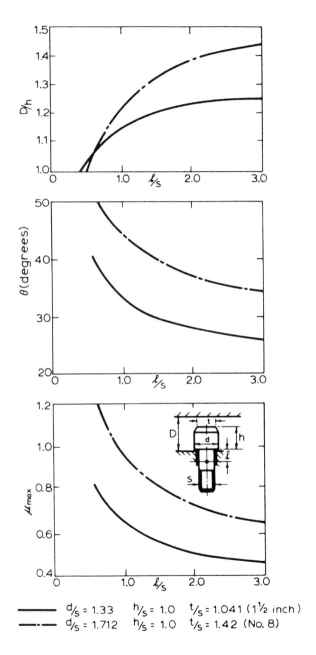

Fig. 5.17 Optimum values of track angle θ and track depth D for maximum coefficient of friction μ_{max}: cap head or hexagon socket screws (from Ref. 3).

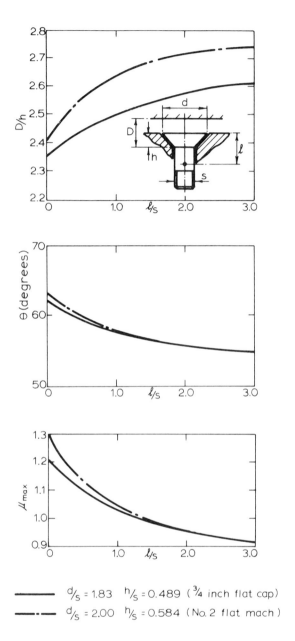

------- $d/_s = 1.83$ $h/_s = 0.489$ (¾ inch flat cap)
-·-·-· $d/_s = 2.00$ $h/_s = 0.584$ (No. 2 flat mach)

Fig. 5.18 Optimum values of track angle θ and track depth D for maximum coefficient of friction μ_{max}: flat-head cap or machine screws in V track (from Ref. 3).

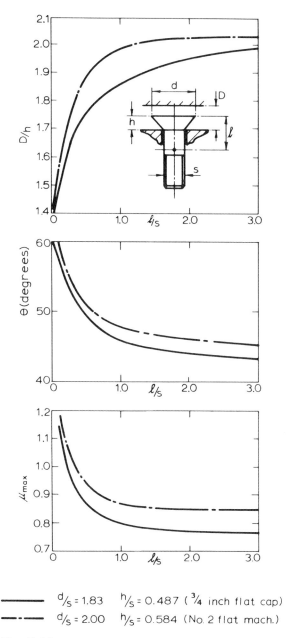

------- $d/s = 1.83$ $h/s = 0.487$ ($3/4$ inch flat cap)

--·-- $d/s = 2.00$ $h/s = 0.584$ (No. 2 flat mach.)

Fig. 5.19 Optimal values of track angle θ and track depth D for maximum coefficient of friction μ_{max}: flat-head cap or machine screws in plain track (from Ref. 3).

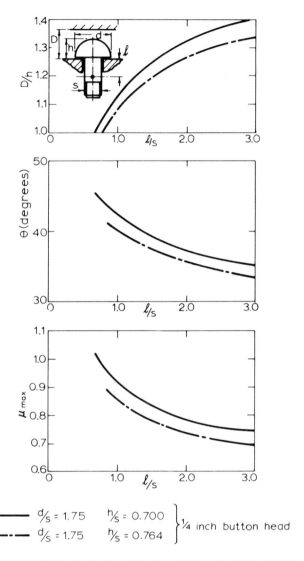

Fig. 5.20 Optimal value of track angle θ and track depth D for maximum coefficient of friction μ_{max}: button-head cap or round-head machine screws (from Ref. 3).

The optimum values of the track angle θ and the ratio of track depth to screw-head depth are then read off the appropriate curves in the figures. In all cases except the cap-head screw of the hexagon socket type (Fig. 5.17), the ranges of these values are quite small. When using this figure, therefore, remember that the lower line is for the large sizes and the upper line is for the small sizes.

Finally, reference to the lower graph will give the maximum permissible value of μ_s. If the actual value of μ_s is greater than this, the screw cannot be fed in a slotted gravity-feed track.

Although the gravity-feed track is the simplest form of feed track, it has some disadvantages. The main disadvantage is the need to have the feeder in an elevated position. This may cause trouble in loading the feeder and in freeing any blockages that may occur. In such cases, the use of a powered track may be considered.

5.2 POWERED FEED TRACKS

The most common types of powered feed tracks are vibratory tracks and air-assisted tracks. A vibratory feed track, illustrated in Fig. 5.21, operates on the same principle as a vibratory-bowl feeder (Chapter 3). With this device, the track is generally horizontal, and its performance is subject to many of the limitations of conventional vibratory-bowl feeders. In the feeder shown, the vibrations normal and parallel to the track are in phase, and the feeding characteristics will be affected by changes in the effective coefficient of friction μ between the parts and the track. This was illustrated in Chapter 3, where it was shown that an increase in μ generally gives an increase in the conveying velocity. For example, a typical operating condition occurs when the normal track acceleration $A_n = 1.1g_n$. In this case, stable feeding takes place, and Fig. 3.8 shows that a change in μ from 0.2 to 0.8 increases conveying velocity from 18 to 55 mm/s when operating at a frequency of 60Hz. It is shown in Appendix B that introducing the appropriate phase difference between the components of vibration normal and parallel to the track can give conditions in which the conveying velocity is consistently high for a wide range of values of

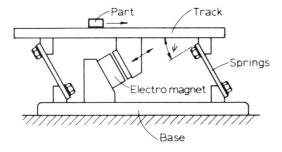

Fig. 5.21 Horizontal vibratory feed track.

μ. With a track inclined at 4 deg (0.07 rad) to the horizontal (a typical figure for a bowl feeder), the optimum phase angle for a normal track acceleration of 1.2 g is approximately -65 deg (1.1 rad). Figure B.2 shows that this optimum value is not significantly affected by the vibration angle employed. In the work that led to these results, it was also found that the optimum phase angle reduces as the track angle is reduced until, when the track is inclined downward at an angle of 8 deg (0.14 rad), the optimum phase angle is almost zero. This means that a drive of the type shown in Fig. 5.21 operates under almost optimum conditions for a track inclined at 8 deg downward. Under these circumstances, it is found that the mean conveying velocity for a wide range of μ is given by

$$v_m = 4500/f \text{ mm/s} \tag{5.41}$$

where f is the frequency of vibration in Hertz when the vibration angle is 20 deg (0.35 rad) and $A_n/g_n = 1.2$. In this case, the mean conveying velocity for an operating frequency of 60 Hz would be 75 mm/s. A higher feed rate could be obtained by reducing the vibration angle, increasing the downward slope of the track, or reducing the frequency of vibration.

For downward-sloping tracks with a large inclination, large feed rates can be obtained with zero vibration angle. In this case, the track is simply vibrated parallel to itself; typical feed characteristics thus obtained are illustrated in Fig. 5.22, which shows the effect of parallel track acceleration on the mean conveying velocity for various track angles when the coefficient of friction is 0.5. The results for a feed track with parallel vibration can be summarized by the empirical equation

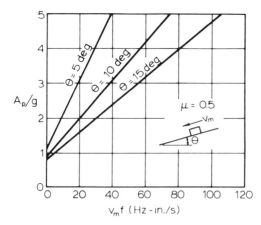

Fig. 5.22 Effect of parallel track acceleration on mean conveying velocity.

$$v_m = \frac{(\theta/f)(A_p/g + 9.25 \times 10^{-4}\theta^2 - 2.3\mu + 0.25)(25.4)}{0.007\theta + 0.07 + 0.8\mu} \quad (5.42)$$

where v_m = mean conveying velocity, mm/s

f = vibration frequency, Hz

θ = inclination of downward sloping track, deg

μ = effective coefficient of friction between part and track

A_p = parallel track acceleration, mm/s^2

g = acceleration due to gravity (9810 mm/s^2)

Equation (5.42) applies when $A_p/g \geqslant 2.0$, $0.3 \leqslant \mu \leqslant 0.8$, $5 \leqslant \theta \leqslant 25$ and when $\theta \leqslant 28.5\mu + 2.5$.

5.2.1 Example

A track inclined downward at 10 deg is vibrated parallel to itself at a frequency of 60 Hz with an amplitude of vibration of 0.2 mm. If the coefficient of friction between the part and the track is 0.5, the mean conveying velocity of the part may be estimated as follows. The dimensionless parallel track acceleration is given by

$$\frac{A_p}{g} = \frac{(0.2)(120\pi)^2}{9810} = 2.9$$

Since $\theta \leqslant 28.5\mu + 2.5$ and $A_p/g > 2.0$, Eq. (5.42) applies and, therefore,

$$v_m = \frac{(10/60)[2.9 + 9.25 \times 10^{-4} \times 10^2 - 2.3(0.5) + 0.25](25.4)}{0.007(10) + 0.07 + 0.8(0.5)} = 16 \text{ mm/s}$$

Air-assisted feed tracks are often simply gravity-feed tracks with air jets suitably placed to assist the transfer of the parts (Fig. 5.23). These devices are

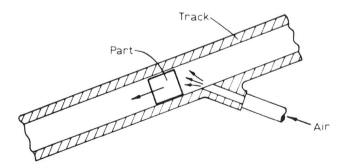

Fig. 5.23 Air-assisted feed track.

ideal for conditions in which the gravity-feed track alone will not quite meet the requirements. Although a well-designed air-assisted feed track can, under suitable conditions, feed parts up an inclined track, it is more usual for the track to slope downward or to be horizontal.

For all types of feed tracks, two important points must be considered. For good utilization of space, a feed track should not be too long. Conversely, the feed track, besides acting as a transfer device, provides a buffer stock of parts which, if a blockage occurs, will allow the workhead to continue operating for a limited period. Ideally, the feed track should be capable of holding enough parts to ensure that the workhead can continue to operate long enough to allow the blockage to be detected and cleared.

A further requirement for all feed tracks, and indeed all parts feeders and workheads, is that, in the event of a blockage, the parts are readily accessible. For this reason, feed tracks should be designed to allow easy access to all parts of the track.

5.3 ESCAPEMENTS

Many types of escapements have been developed and, quite often, for a given part, several available types will perform the required function. Figure 5.24 shows two examples of what is probably the simplest type of escapement. Here the parts are pulled from the feed track by the work carrier, and the escapement itself consists of only a rocker arm or a spring blade.

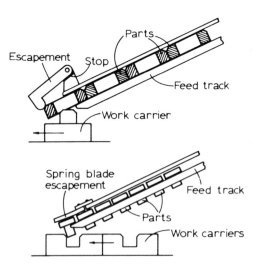

Fig. 5.24 Escapements actuated by the work carrier.

Many escapements are not always recognized as such. For example, a rotary indexing table may be arranged to act as an escapement. This is shown in Fig. 5.25, where parts may be taken from either a horizontal-delivery feed track or a vertical-delivery feed track.

An advantage of the simple escapements illustrated in Figs. 5.24 and 5.25 is that they also act as parts-placing mechanisms. The two escapements shown in Fig. 5.24 are somewhat unusual in that they are activated by the work carrier and part whereas, with most escapements, the motion of the part is activated by the escapement which, in turn, is activated by some workhead function. This latter method is the most common in practice, and escapements of this type may be subdivided into various categories, as described below.

5.3.1 Ratchet Escapements

Two examples of ratchet escapements are shown in Fig. 5.26; the functions being performed are different in each case. The pawl is designed so that, as its front finger lifts clear of the line of parts, its back finger retains either the next part, as shown in Fig. 5.26a, or a part further up the line, as shown in Fig. 5.26b. Ratchet escapements operating on several feed tracks can be activated from a single mechanism. In automatic assembly, the release of several parts from a single feed track is not often required, but the release of one part from each of several feed tracks is often desirable. This can be achieved by a series of ratchet escapements of the type shown in Fig. 5.26a. As the escapement is activated, the front finger rotates but without moving the part to be released. At position 2, the front finger is just about to release part *A*, and the back finger moves in such a way that no motion is imparted to part *B*. On the return

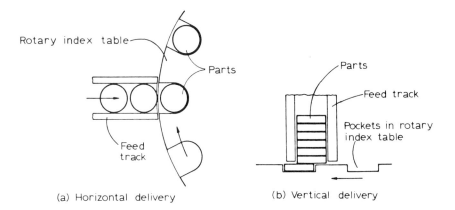

(a) Horizontal delivery　　　　　(b) Vertical delivery

Fig. 5.25 Feeding of parts onto rotary index table.

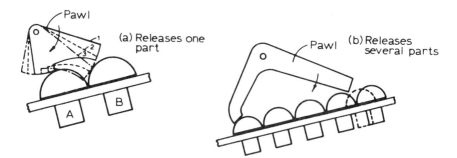

Fig. 5.26 Rachet escapements operated by rotary motion.

stroke of the escapement, part B is released by the back finger and retained by the front finger. Figure 5.26b shows a similar type of mechanism except that the remaining parts in that track move forward before being retained by the back finger. In these devices, the back finger of the escapement should not produce motion of the parts opposite to the direction of flow. If this tendency is present, either all the parts above this point on the feed track will move backward and may subject the escapement to heavy loads or the parts may lock and cause damage to the escapement.

In the examples already shown, the motions of the front and back fingers of the escapement are obtained by a rotary motion. The fingers of a ratchet escapement may, however, be operated together or independently by cams, solenoids, or pneumatic cylinders, giving a linear motion as shown in Fig. 5.27.

It is clear from all the foregoing examples of ratchet escapements that the escapement can be used only to regulate the flow of parts which, when arranged in single file, have suitable gaps between their outer edges.

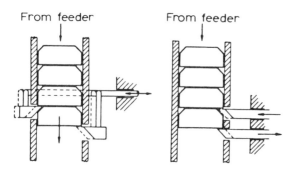

Fig. 5.27 Ratchet escapements operated by linear motion.

5.3.2 Slide Escapements

Five examples of slide escapements are shown in Figs. 5.28–5.30. It can be seen from the figures that, in the slide escapement, one or more parts are removed from the feed chute by the action of a cross slide and that applications of this type of device are restricted to parts that do not interlock with each other. The slide escapement is ideally suited to regulating the flow of spherical, cylindrical, or platelike parts and although, in Figs. 5.28–5.30, the feed track enters the escapement vertically, this is desirable but not necessary.

As with the ratchet escapement, parts may be released either singly or in batches from one or a number of feed tracks by the action of a single actuating mechanism. A further alternative, not available with the ratchet escapement, is for parts fed from a single feed track to be equally divided between two

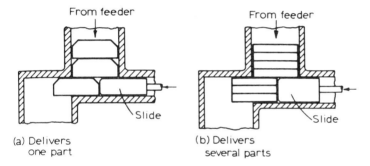

(a) Delivers
    one part

(b) Delivers
    several parts

Fig. 5.28 Slide escapements delivering into single feed chute.

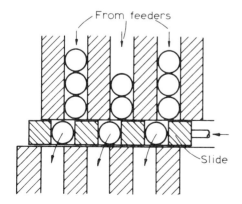

Fig. 5.29 Slide escapements operating several feed chutes.

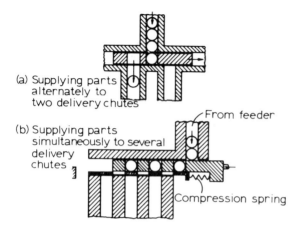

(a) Supplying parts alternately to two delivery chutes

From feeder

(b) Supplying parts simultaneously to several delivery chutes

Compression spring

Fig. 5.30 Slide escapements supplying two or more delivery chutes from a single feed chute.

delivery tracks, as shown in Fig. 5.30a. This type of escapement is very useful where two identical parts are to be used in equal quantities and a parts feeder is available that will deliver at a sufficient rate to meet the total requirement. If it is necessary to feed parts from a single feed track into more than two delivery tracks, a slide escapement of the type shown in Fig. 5.30b is suitable. The figure shows three delivery tracks being fed from a single feed track and, as before, only one actuator is necessary.

5.3.3 Drum Escapements

Two types of drum escapements, usually referred to as *drum-spider* escapements, are shown in Fig. 5.31 where, in these cases, the drum is mounted vertically and the parts are either fed and delivered side by side (Fig. 5.31a) or fed end to end and delivered side by side (Fig. 5.31b). In the latter case, the parts are fed horizontally to the escapement. One advantage of the vertical drum escapement is that a change in the direction of motion of the parts is easily accomplished. This can be very useful where the horizontal distance between the parts feeder and the workhead is restricted to a value that would necessitate very sharp curves in the feed track if an alternative type of escapement is used. A further feature of the drum-spider escapement is that the parts on passing through the escapement, can be turned through a given angle.

Two other types of drum escapements, the *star wheel* and the *worm*, are shown in Fig. 5.32, where it can be seen that the direction of motion of the parts is unaffected by the escapement and the parts must have a suitable gap between their outer edges when they are arranged in single file.

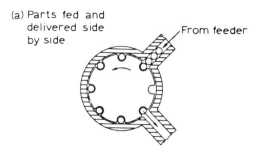

(a) Parts fed and delivered side by side

From feeder

(b) Parts fed end to end and delivered side by side

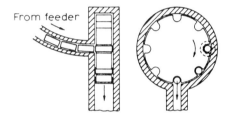

From feeder

Fig. 5.31 Drum-spider escapements.

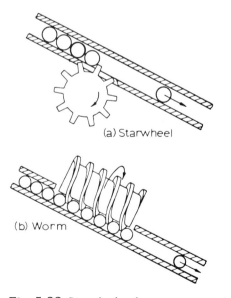

(a) Starwheel

(b) Worm

Fig. 5.32 Star-wheel and worm escapements.

Drum escapements may be driven continuously or indexed, but usually an indexing mechanism is preferable because difficulties may be encountered in attempts to synchronize a continuous drive to meet the requirements of the workhead. Of the six types of drum escapements described (Figs. 5.30–5.32), the first four are effectively rotary slide escapements, and the latter two are rotary ratchet escapements.

5.3.4 Gate Escapements

The gate escapement is seldom used as a means of regulating the flow of parts to an automatic workhead on a mechanized assembly machine. Its main use is to provide an alternative path for parts and, in this capacity, it is often used for removing faulty parts from the main flow. However, one type of gate escapement, shown in Fig. 5.33, can be used to advantage on certain types of parts when it is necessary to provide two equal outputs from a single feed track input. It is clear from the figure that although this device is usually referred to as an escapement, it does not regulate the flow of parts, and further escapements would be necessary on the delivery tracks to carry out this function.

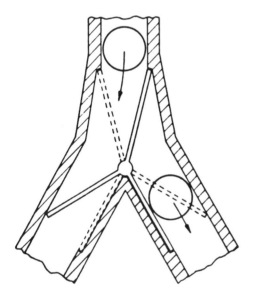

Fig. 5.33 Gate escapement.

5.3.5 Jaw Escapements

Jaw escapements are particularly useful in automatic assembly applications in which some forming process is necessary on the part after it has been placed in position in the assembly. Figure 5.34a shows an example of such an application. The part is held by the jaws until the actuator, in this case a punch, forces the part through the jaws. The punch then performs the punching operation, returns, and allows another part into the jaws. The device can, of course, be used purely as an escapement.

5.4 PARTS-PLACING MECHANISMS

Two simple types of parts-placing mechanisms have already been described and illustrated in Figs. 5.24 and 5.25. In the first two examples, the parts are taken from the feed track by the work carrier and, in the other cases, the parts are fed by gravity into pockets on a rotary index table. These special applications, however, can be used only for a limited range of parts, and by far the most widely used parts-placing mechanism is a conventional gravity-feed track working in conjunction with an escapement. This system of parts placing is probably the cheapest available, but it has certain limitations. First, it may not be possible to place and fasten parts at the same position on the machine because of interference between the feed track and the workhead. This would necessitate a separate workstation for positioning the part, which would result in an increase in the length of the machine. It then becomes necessary to retain

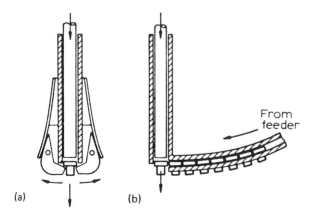

Fig. 5.34 (a) Jaw escapement and (b) assisted gravity-feed part-placing mechanism.

the part in its correct orientation in the assembly during transfer. Second, if a close fit is required between the part and the assembly, the force due to gravity may not be sufficient to ensure that the part seats properly. Third, if the part cannot be suitably chamfered, the gravity-feed track may not give the precise location required. However, for placing screws and rivets prior to fastening, which forms a large proportion of all parts-placing requirements, and where the tool activates the escapement and applies the required force to make assembly possible, the gravity-feed track is invariably used. An example of automatic screw placing and driving is shown in Fig. 5.35.

A further example of the assisted gravity-feed type is shown in Fig. 5.34b, where a part is being positioned in the assembly. With this device, the feed track positions the part vertically above the assembly. The part is then guided into position in the assembly by a reciprocating guide rod. With these systems, it is common to fit the escapement at the point at which the guide rod operates and to use the guide rod as the escapement activator. In some applications, the part may be positioned above the assembly by means of a slide escapement and then guided into position using a guide rod. This system is commonly referred to as the *push-and-guide system* of parts placing.

For situations in which the placing mechanism has to be displaced from the workstation location, the *pick-and-place system* is often used. The basic action of this system is shown in Figs. 5.36 and 5.37, where it can be seen that the part is picked up from the feed track by means of a mechanical, magnetic, or vacuum hand, depending on the particular application, placed in position in the assembly, and then released. The transfer arm then returns along the same path to its initial position. Some units pick vertically, transfer along a straight

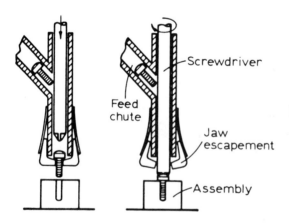

Fig. 5.35 Parts-placing mechanism for automatic screwdriver.

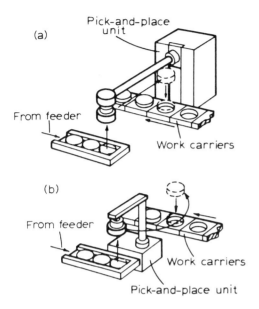

Fig. 5.36 Pick-and-place units that lift and position vertically.

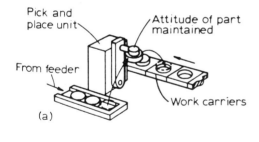

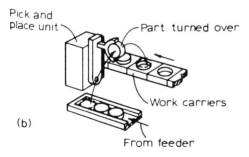

Fig. 5.37 Pick-and-place units that move the part along the arc of a circle.

path horizontally, and place vertically, as shown in Fig. 5.36a. Others pick vertically, transfer around the arc of a circle in a horizontal plane, and place vertically, as shown in Fig. 5.36b. A third type picks and places by rotary motion of the transfer arm in a vertical plane and, throughout transfer, the hand remains vertical (Fig. 5.37a). Finally, a variation of this latter type has the pickup head fixed to the arm, and the part is turned over during the operation (Fig. 5.37b). When operating correctly, the first two examples can mate parts that have close assembly tolerances and no chamfers, whereas the latter two examples, although the simplest systems, can operate only when the length of the vertical contact between the mating parts is small.

On high-speed operations, the pick-and-place mechanisms are usually mechanically actuated and are mechanically connected to the entire assembly operation. Each unit is designed and built for a particular application. However, because of the similarity between many of the assembly operations, modular pick-and-place mechanisms have been developed. These basically off-the-shelf units can handle a large percentage of assembly operations with little modification. These units, which were developed in-house by assembly machine manufacturers as part of their basic assembly machinery, have been made available by smaller manufacturers as separate items. The modular concept has been successful because many assembly operations consist of the same sequence of simple motions.

Recent studies have indicated that most subassemblies and some complete assemblies can be put together with a single pick-and-place unit with three or four degrees of freedom, as represented by a rectangular or cylindrical coordinate system with a twisting motion about the major axis. Such a mechanism needs an elaborate control system to handle the larger variety of sequential motions. These systems, including the mechanisms and controls, are called *robots* and can replace the dedicated pick-and-place mechanisms.

5.5 ASSEMBLY ROBOTS

Since the 1970s, great interest has been shown in the use of robots in assembly automation. High-speed automatic assembly is generally applicable only to high-volume production (at least 1 or 2 million per year) and in situations in which the product design is stable and a steady demand is anticipated. In the early 1970s, it was felt that the introduction of general-purpose assembly robots would allow the application of automation to the assembly of products that did not meet these high-volume production requirements. The dream was to develop assembly systems so flexible that they could be quickly adapted and reprogrammed to assemble different products in medium, or even small, batches. Companies purchased assembly robots in order to test them on their own products, but the results were disappointing. Unfortunately, in the United

States, interest concentrated on the high-technology aspect of robotics—including vision systems, tactile sensing, and expensive multi-degree-of-freedom robots expected to operate at a single workstation and perform the tasks of a human assembly worker. Indeed, one of the more popular U.S. assembly robots, the PUMA, was designed to have reach and dexterity characteristics similar to those of a human being.

In the meantime, in Japan, a more practical approach was being used in which robots were treated simply as another tool to be applied, where suitable, to assist in assembly work.

Before examining this situation in greater detail, let's clarify some of the terminology being used and then look at some of the advantages of assembly robots.

5.5.1 Terminology

Assembly automation equipment used in mass production is often referred to as *dedicated* equipment. Assembly automation of the more flexible kind is referred to as *programmable* or *adaptable assembly* or just *robot assembly*. Unfortunately, the use of these terms sometimes leads to considerable confusion. For example, a multistation assembly machine for producing one tape deck mechanism every few seconds would be regarded as a dedicated machine. However, this machine could be directed by a computer to skip assembly stations on command and thereby produce various styles of the tape deck. We would now have a programmable machine, and yet no assembly robots would be involved. On the other hand, an assembly system could be developed that used one or more assembly robots to assemble the tape decks. The machine could still be dedicated to the one product, and yet we now have programmable or robot assembly.

In fact, almost all robot assembly applications in present practice use the robots as dedicated equipment, and the basic programming of the robot might be done only during initial set up of the system, with occasional later reprogramming to allow for design changes or style variations. Hence, it is misleading and confusing to speak of robot assembly equipment or programmable assembly equipment as the alternative to dedicated assembly equipment.

In this book, the term *special-purpose* is used to describe equipment or tooling that is custom-made for one product or part and the term *general-purpose* is used to describe equipment or tooling that can be readily adapted or programmed for a variety of products or parts.

5.5.2 Advantages of Robot Assembly

Some of the main advantages in the use of assembly robots can be described with reference to the conditions for the economic application of special-purpose assembly machines.

1.   Stability of the product design. If the product design changes, the robot can be reprogrammed accordingly. However, this does not usually apply to the peripheral items in the system that contact the parts, such as feeders, grippers, etc.

2.   Production volume. As will be seen, a robot system can operate economically at much longer station cycle times than a high-speed automatic assembly machine. (The station cycle time is the interval between the production of completed assemblies.)

3.   Style variations. A robot system can more readily be arranged to accommodate various styles of the same product. For example, the robot can be reprogrammed to select only certain parts for the assembly, depending on the particular style required.

4.   Part defects. First, it is interesting to note that a feeder jam caused by a faulty part causes much greater loss in production on a high-speed transfer assembly machine than on a robot system with a relatively long cycle time. In addition, the robot can be programmed to sense problems that may occur and to reattempt the insertion procedure.

5.   Part size. As will be seen later, a principal advantage of a robot used in assembly is that the parts can be presented in patterns or arrays on pallets or part trays. In this case, the severe restrictions on part size in high-speed automation do not apply.

In the automatic assembly of one part, there are two principal steps: (1) the handling and presentation of the part to the insertion device and, (2) the insertion of the part. Depending on circumstances, either or both of these steps might be carried out automatically, and the last step (insertion) might be carried out by an assembly robot. It is useful to consider the alternatives for the automation of the insertion operation, assuming first that the parts can be presented at the required frequency and in the same orientation. First, robot arms can be employed. However, it should be noted that there are some insertion processes a robot usually cannot perform without the aid of special workstations; these would include, for example, all insertion or fastening operations requiring the application of large forces. Also, there are some processes that a robot cannot perform without using special tools; these would include screw insertion, spot-welding, etc. For many of these cases, a single-part special-purpose workhead can be used that is a mechanism or machine designed to perform the insertion operation repeatedly.

On the other hand, there are some operations that require such complicated manipulations that a robot under program control is essential. These include following a contoured seam in welding or running a wire through a complicated path. The more fundamental difference between a robot and a single-part special-purpose workhead, however, is that the workhead can perform only the one operation repeatedly, whereas the robot can perform a sequence of different operations repeatedly.

When the required assembly rate is high, a transfer machine is required and, at each station, the same operations must be performed repeatedly. Under these circumstances, the decision to use a special-purpose workhead or a robot at a particular station depends on the nature of the insertion operation and on the cost of the equipment. By the time a special-purpose workhead has been adapted and engineered for an assembly machine and then debugged, its final cost may be double the original cost. One advantage of a robot is the reduction in the cost of adaptation.

It would appear that, from the viewpoint of the economics of assembly processes, the main distinguishing features of a robot are that it can be programmed to perform a cycle of several different operations quickly and can be engineered easily to adapt to changes in product design or style variations.

In this book, it has been seen that automatic parts presentation is one of the major problems in high-speed automatic assembly. Parts must be presented to a single-part, special-purpose workhead in the same orientation at the same location. Vibratory-bowl feeders and a variety of mechanical feeders can be economically used for this purpose as long as the part is small and is required at relatively frequent intervals. These conditions arise from the fact that the cost of automatically feeding a part is proportional to the length of the assembly cycle.

For example, if it is assumed that simple parts 25 mm long can be fed at 1/sec at a cost of 0.04 cents each using a vibratory-bowl feeder, then, with the same bowl feeder, a part 250 mm long would take 10 sec to deliver at a cost of 0.4 cents. Since it would cost about 0.4 cents for an assembly worker to perform the operation, it would not be economical to feed larger parts automatically. Correspondingly, it would not be economical to employ a vibratory feeder if the assembly cycle is relatively long. It is assumed here that the part does not present particular feeding and orienting problems because of geometry, flexibility, or fragility, etc. If any of these problems exist, then of course, the economics of automatic handling would be even less attractive.

Another method of parts presentation is by magazine. Sometimes the parts can be purchased in connected form like staples or secured to paper strips and then coiled, as with many electronic components. Under these circumstances, the orientation has been carried out by the manufacturer, and this makes parts presentation in automatic assembly relatively inexpensive.

Under some conditions, it is economical to have a special magazine associated with the special-purpose workhead and load this with parts manually. In this case, the part orientation is done manually, either at the machine or "off-line." If the loading is done off-line, there is the possibility that the magazines can be loaded automatically, sometimes immediately after part manufacture. This situation lends itself to preinspection to avoid assembly machine downtime due to faulty parts.

When assembly robots are employed, further important possibilities exist for parts presentation:

1. Some parts may be presented partially oriented, and the robot can perform final orientation,
2. Identical parts may be presented in pallets or part trays in fixed arrays,
3. Sets of different parts can be presented in part trays (kits),
4. Feeders might be used that can feed different parts simultaneously.

We can conclude that a further advantage of the use of assembly robots is the widely increased alternatives allowed in the methods of parts presentation. Some of these alternatives will now be discussed in more detail.

5.5.3 Magazines

Part magazines provide one of the more convenient ways to present parts to the assembly system and have been studied by Arnstrom and Grondahl [4]. The purposes of a magazine are two fold: (1) to present the parts in the same orientation to the robot, and (2) to decouple the manual handling process from the machine cycle time.

If both these requirements are met, the assembly system is able to operate automatically. Often, the requirements can be met with simple and inexpensive methods. Sometimes, however, questions concerning space requirements, precision, transportation, storage, and cost must be carefully considered. Magazines can perform the following additional functions:

1. If part manufacture and assembly can be integrated, the magazine can form a buffer between the two operations. This can help to balance fluctuations in cycle time and minimize imbalances between manufacturing and assembly requirements.

2. The magazine can provide a method for holding the parts during transportation and so rationalize transportation procedures as well as minimize the risk of damage to the parts.

3. Magazines can help to simplify the problems of part storage and inventory control.

It is important to make a distinction between a magazine and a buffer. The objective in using a magazine is primarily to decouple the operator from the machine to avoid costly idle time when the operator has to wait for the machine or the machine has to wait for the operator.

The magazine coupled directly to the assembly system creates opportunities for the operator to attend to other aspects of machine loading. The function of a buffer is to decouple one machine station from the adjacent station. The buffer accommodates the idle times that would occur if one station had to wait for another. A buffer eliminates varying imbalances between stations on a free-transfer, multistation machine.

5.5.4 Types of Magazine Systems

Figure 5.38 presents a rough classification of different magazine configurations. This classification is based on the requirements imposed by the part-presentation mode on the robot motions. It should be noted that the part-presentation mode also influences the space requirements for the magazine. It can be seen that magazines can be divided into two basic types:

1. Part presentation at a single location (Fig. 5.38a). With these magazines, the robot acquires the part always at the same location. The parts move within the magazine and are fed by gravity or a driving force, such as that exerted by a spring, for example, each time a part is removed from the magazine.

2. Part presentation at multiple locations (Fig. 5.38b). In these magazines, the parts are placed in a fixed array. The robot must be programmed to acquire the parts in a predetermined pattern (pattern picking).

5.5.5 Automatic Feeders for Robot Assembly

Numerous developments in automatic parts feeding for robot assembly are taking place. These range from simple mechanically adjustable feeders to sophisticated "vision" systems. Since it is not possible in this paper to review all the different approaches, only a few representative examples will be mentioned. Examples will be limited to those that seem to hold both practical and economic possibilities.

One important objective in any feeder suitable for robot assembly is to minimize the total cost of the feeder divided by the number of different parts to be fed. Several so-called multipart feeders have been under development, some employing vision systems such as the Hitachi feeder, [5] and at least one in which interchangeable mechanical tooling is used, such as, the so-called Salford feeder [6].

In the Hitachi feeder, several vibratory bowls (one for each different part) feed the parts without orientation to a vision station. Each part is held momentarily in the station, where a camera determines which of the limited number of possible stable orientations exist. This information then allows the robot to grip the part correctly and perform the necessary orientation when transferring

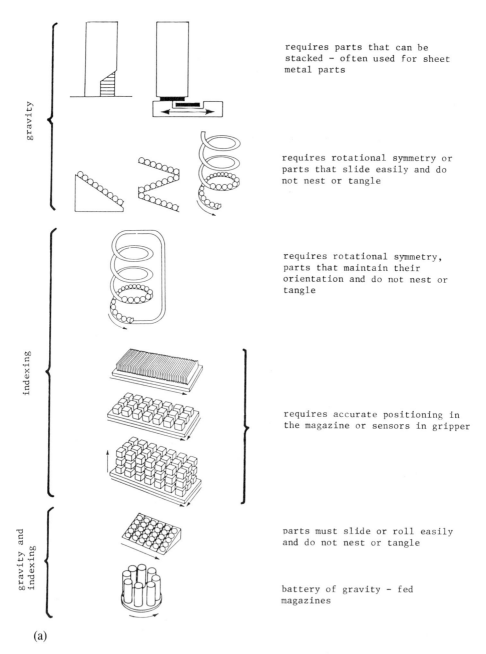

(a)

Fig. 5.38 (a) Part magazine configurations: part presentation at a single location; (b) Part magazine configurations: part presentation at multiple locations (from Ref. 4).

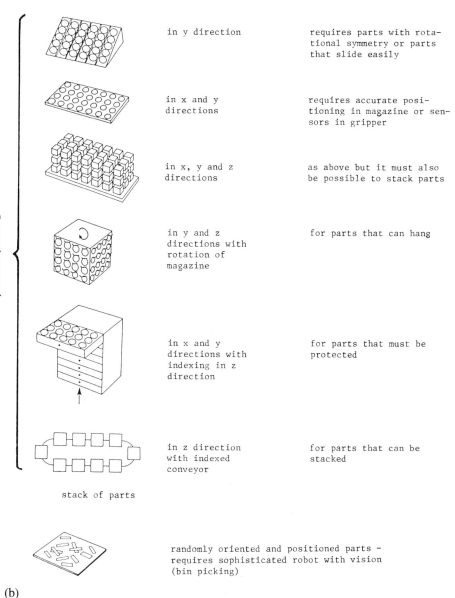

in y direction   requires parts with rotational symmetry or parts that slide easily

in x and y directions   requires accurate positioning in magazine or sensors in gripper

in x, y and z directions   as above but it must also be possible to stack parts

in y and z directions with rotation of magazine   for parts that can hang

in x and y directions with indexing in z direction   for parts that must be protected

in z direction with indexed conveyor   for parts that can be stacked

stack of parts

randomly oriented and positioned parts – requires sophisticated robot with vision (bin picking)

(b)

it from the feeder to the assembly. Such an arrangement allows the relatively large cost of the equipment to be distributed among several parts. In addition, the amount of special-purpose equipment is kept to a minimum.

The technique of employing a simple vision system to distinguish between a limited number of part orientations in this way was first developed by Heginbotham et al. [7] and later applied to a multipart feeder [5].

The Salford multipart feeder [6] consists of a linear vibratory drive onto which several different straight feed tracks can be mounted, with provision for returning rejected parts to the beginning of each track using a reciprocating elevator. The feed tracks are machined especially for each part. The cost of each track is kept to a minimum by using a library of programs for an N.C. milling machine. The orienting devices that can be machined in this way are limited in type so that the feeder is not as versatile as a vibratory-bowl feeder. Nevertheless, the total cost of the feeder per part type is likely to be very low, making it a most attractive proposition in suitable circumstances. One disadvantage with multipart feeders is that they will not necessarily be applicable in multistation robot systems. Such a feeder can be used only when a significant number of parts being assembled at a particular station can be fed by that feeder.

Single-part feeders for robot assembly (Fig. 5.39) have been developed [8]. These are based on the double-belt principle and, for many simple rotational parts, require only one interchangeable blade to perform all the necessary orientations. For rotational parts with asymmetric grooves or holes, the final orientation can be performed by the robot, which is programmed to rotate the part above a single sensor until the feature is detected. This operation leads to an increase (of about 1 or 2 sec) in the station cycle time.

For nonrotational parts, the same feeder can be fitted with a single sensor and a solenoid-activated rejection device. Such an arrangement is capable of handling a very wide range of parts.

One particularly useful attribute of these double-belt feeders is that no special-purpose delivery track is required, and special-purpose tooling costs can be kept to a minimum.

5.5.6 Economics of Part-Presentation

A study [9] has been made of the comparative economics of a range of part-presentation devices. The study was based on the idea that the cost of any part-presentation system per part type can be divided into three components:

1. The basic cost of the feeder (general-purpose equipment)
2. The cost of the tooling that is special to the part under consideration (special-purpose equipment)
3. The cost of manual labor

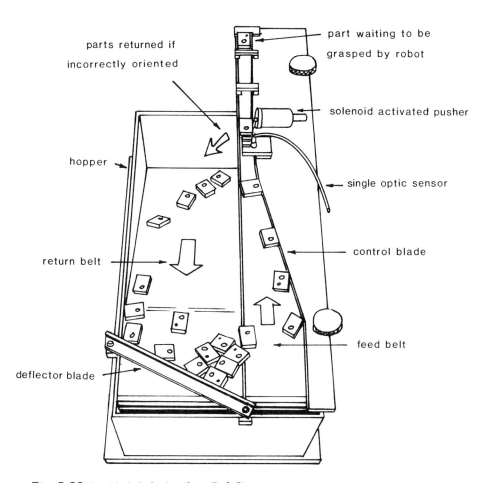

Fig. 5.39 Double-belt feeder (from Ref. 8).

In multistation machines, it is possible to interpose manual assembly stations between automatic stations whereas, in single-station robot systems, it is not usually feasible, for safety considerations, to arrange for manual operations within the system cycle. For those situations in which manual insertion of parts is a possible alternative, the cost C_{mi} for the manual insertion of one part would be given by

$$C_{mi} = (t_{at}/n_a)W_a \text{ cents} \qquad (5.43)$$

where t_{at} is the machine cycle time (sec), n_a the number of parts manually assembled consecutively during one machine cycle, and W_a the assembly worker rate (cents/sec).

Normally, n_a will be unity and, since a typical average manual assembly time is 8.25 sec [10], Eq. (5.43) is valid only for values of t_{at} greater than this. However, in some multistation free-transfer machines using robots, the station cycle time may be long enough to allow two or three operations at one manual station. In these cases, n_a in Eq. (5.43) would be increased and the minimum value of t_{at} would increase proportionally. Rather than present the costs of manual insertion as a step function, we will assume that these are given by

$$C_{mi} = 8.25\, W_a \text{ cents} \tag{5.44}$$

Equipment is needed for all other part-presentation methods, and the calculations of presentation costs depend on the size of the product batch. When the batch is small, the cost attributable to special-purpose tooling is given by distributing the equipment cost over the batch. The cost attributable to equipment that can be reused is obtained by depreciating the equipment cost over the time worked during the equipment payback period. Thus, in general, for small batches, the cost of automatic part presentation is given by

$$C_f = C_{gp}(0.014/P_s)\, t_{at} + 100 C_{sp}/B_s \text{ cents} \tag{5.45}$$

$$
\begin{array}{ccc}
\text{Cost of} & \text{Cost of using general} & \text{Cost of using} \\
\text{part} = & \text{purpose equipment} + & \text{special-purpose} \\
\text{feeding} & & \text{equipment}
\end{array}
$$

where C_{gp} and C_{sp} are general-purpose and special-purpose equipment costs, respectively (in thousands of dollars), P_s is the number of shift-years in the payback period, and B_s is the size of the batch (in thousands) to be produced during the equipment payback period. It should be noted that the constant 0.014 is a combined factor to convert shift-years to seconds and dollars to cents. If the time taken to manufacture the complete batch $(t_{at} B_s)$ is greater than the equipment payback period P_s, then Eq. (5.45) becomes

$$C_f = (C_{gp} + C_{sp})(0.014/P_s)\, t_{at} \text{ cents} \tag{5.46}$$

For manually loaded magazines, the general-purpose portion of the equipment cost will be small, but an extra term must be added to allow for the cost of manual handling for loading each part into the magazine. This cost is given by multiplying the manual assembly time t_m by the assembly worker rate W_a and is estimated to be approximately 0.7 cents [10]. Thus, the cost of part presentation by manually loaded magazine C_{mm} becomes

$$C_{mm} = 100\, C_{sp}/B_s + t_m W_a \text{ cents}$$

when $t_{at} B_s < 7200\, P_s$ and

$$C_{mm} = C_{sp}(0.014/P_s)t_{at} + t_m W_a \text{ cents} \tag{5.47}$$

when $t_{at} B_s > 7200\, P_s$

Equations (5.44–5.47) can be used to determine, for any part-presentation method, the conditions for which the method would be economic compared with manual insertion or manually loaded magazines. Figure 5.40 shows the results obtained in this way for the representative part-presentation systems given in Table 5.4. Superimposed on the figure are lines showing values of the total batch production time, a time given by $t_{at} B_s$.

It can be seen that when the total batch production time becomes greater than the equipment payback period, the choice of feeding system becomes independent of batch size; further, Eq. (5.46) shows that it is now the total feeder cost that is important. Under these circumstances, the most attractive presentation methods compared with manually loaded magazines would be the Salford, double-belt, vibratory-bowl, and Hitachi feeders, in that order.

When the batch size is reduced, the vibratory-bowl feeder rapidly becomes uneconomical. For batch sizes less than 35,000, the manually loaded magazines are no longer attractive, and manual insertion is preferable. However,

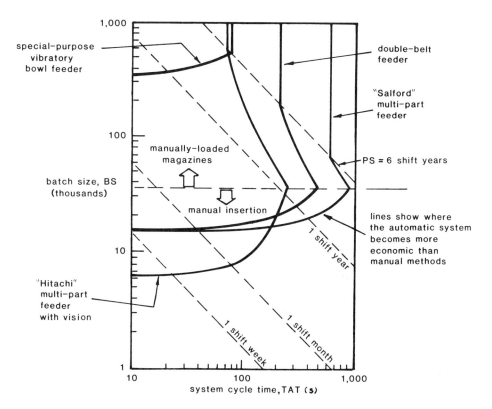

Fig. 5.40 Economics of some representative part-presentation methods (from Ref. 9).

Table 5.4   Costs per Part Type for Some Typical Part-Presentation Systems (Costs Estimated by Author)

Name of System	General-purpose tooling (C_{gp}), in thousands of dollars	Special-purpose tooling (C_{sp}), in thousands of dollars	Labor cost per part
Vibratory-bowl feeder	1.5	3.5	—
Hitachi multipart feeder	5.0	0.2	—
Salford multipart feeder	1.0	0.5	—
Single-part belt feeder	2.0	0.5	—
Manually loaded magazine or pallet	—	1.0	$t_m W_a$ [a]

[a] t_m = time taken to manually load 1 part into magazine or pallet; W_a = operator rate (cents/s)

the three programmable feeder types are still competitive, the Hitachi feeder being the best alternative for batch sizes as low as 6500.

The results in Fig. 5.40 must be interpreted with care since all the alternatives will rarely be suitable for a particular part.

5.5.7 Design of Robot Assembly Systems

The basic configurations of general-purpose assembly robots now being marketed are illustrated in Fig. 5.41. Most of these robots are fixed to a base or pedestal, but a significant number employ robot arms suspended from a gantry or portal framework. It should be pointed out that some automatic assembly systems employ one or more insertion devices fixed above an $x - y$ table such as the Sony FX1 system [11] shown in Fig. 5.42 or systems for assembling printed circuit boards. Although these systems play an important role in automatic assembly, they will not be considered in the present discussion since they do not involve the application of general-purpose assembly robots.

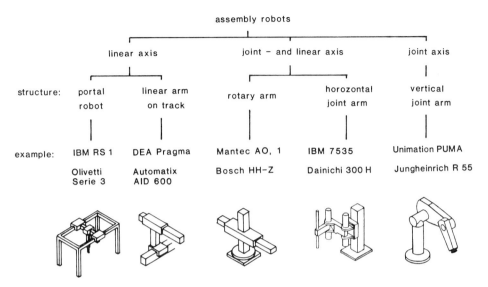

```
                          assembly robots

          linear axis        joint - and linear axis       joint axis

structure:  portal   linear arm   rotary arm   horozontal   vertical
            robot    on track                  joint arm    joint arm

example:   IBM RS 1  DEA Pragma   Mantec AO, 1  IBM 7535    Unimation PUMA

           Olivetti  Automatix    Bosch HH-Z   Dainichi 300 H  Jungheinrich R 55
           Serie 3   AID 600
```

Fig. 5.41 Configurations of general-purpose assembly robots (courtesy Prof. Warnecke, University of Stuttgart).

Assembly robots of the pedestal type can be applied either singly or grouped around one or more stationary work fixtures; they can also be arranged along transfer machines, where each robot will perform several assembly tasks. In either case, the robots can reach into a feeder or pallet to grasp parts.

An example of the single-station robot assembly system has been described by Warnecke and Walther [12]. In this system (Fig. 5.43), two robots work at one station to perform some final assembly tasks on automobile engines. One robot performs the part-handling functions while the other performs the screwing operations. In this way, undesirable gripper or tool changing is avoided. The assembly tasks involve the addition of four subassemblies or parts and fastening with ten screws. Thus, including the engine assembly itself, 15 subassemblies or parts are handled at the station. The station cycle time is 90 sec (6 sec/part) and the manual assembly time for the same tasks is 3.4 min (13.6 sec/part), giving a payback period of 2.5–3 years.

Taniguchi [11] provides an example of the multistation robot assembly system. This system, developed by the Toshiba Corporation, assembles 24 different styles of electric fans and is based on a free-transfer machine with 24 stations (Fig. 5.44). It has 11 general-purpose robots, 2 special-purpose robots, 2 pick-and-place mechanisms, 2 mechanisms for turning over the assembly, 3 special-purpose workheads, and 5 assembly workers.

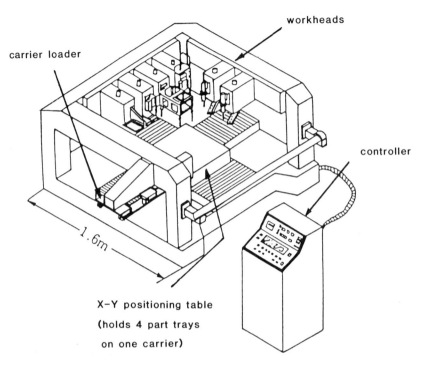

Fig. 5.42 Sony FX-1 flexible assembly system (from Ref. 11).

Both of these systems employ pedestal-type robots. Gantry-type robots are slightly more limited in their applications because they are less convenient to apply in transfer line situations where the series of work carriers must pass under the gantry. However, many system developers consider this arrangement quite feasible, and they envision trays of parts being passed from robot to robot by means of a conveyor. Because of the limited space available within the framework of a gantry-type robot, the parts must usually be conveyed into the working space by delivery tracks. Feeders are now being developed that present the oriented parts within the feeder, thus avoiding the need for special-purpose delivery tracks. Since the gantry robot cannot reach outside its framework, such feeders would have to be contained within the framework which would result in severe space problems.

From the preceding discussion, it can reasonably be assumed for the present purposes that the application of gantry-type robots in assembly is generally limited to those situations in which one or more assemblies are simultaneously built up on stationary work fixtures. Further, most of the parts must

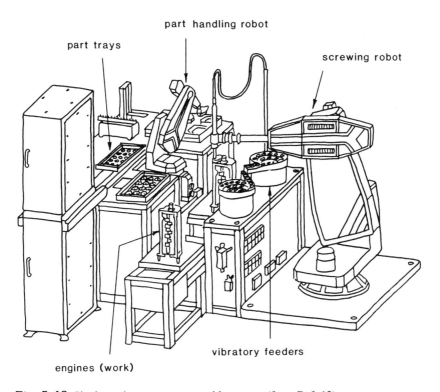

part handling robot

part trays

screwing robot

engines (work)

vibratory feeders

Fig. 5.43 Single-station two-arm assembly system (from Ref. 12).

be presented either on pallets or part trays or in feed tracks or magazines loaded manually or automatically off-line.

The purpose of this discussion, however, is not to attempt to determine which is the "best" assembly robot. In the future, important applications will almost certainly be found for all the various types being developed. It would be useful, however, to determine the production circumstances under which various products would lend themselves to the different configurations of complete assembly systems.

Experience shows that there are important fundamental differences between those assembly systems based on a work-carrier transfer system and those nontransfer systems in which the assembly space is restricted. First, most assemblies include parts that must be manually inserted. In the nontransfer systems, manual operations cannot effectively be carried out within the assembly cycle; manual operations can be done only before or after the robot per-

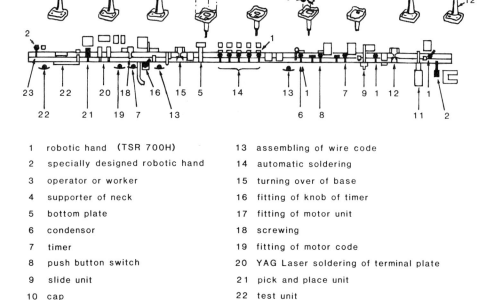

1	robotic hand (TSR 700H)	13	assembling of wire code
2	specially designed robotic hand	14	automatic soldering
3	operator or worker	15	turning over of base
4	supporter of neck	16	fitting of knob of timer
5	bottom plate	17	fitting of motor unit
6	condensor	18	screwing
7	timer	19	fitting of motor code
8	push button switch	20	YAG Laser soldering of terminal plate
9	slide unit	21	pick and place unit
10	cap	22	test unit
11	decorative plate	23	ejection unit
12	base of electric fan		

Fig. 5.44 Multistation free-transfer machine employing assembly robots (from Ref. 11).

forms the consecutive series of automatic operations. This restriction does not apply to the transfer system because assembly workers can be included at any station along the line. For a particular product, this can have an important influence on both the feasibility of robot assembly and the design of the product.

Second, with the nontransfer systems, there is a direct relation between the cycle time and the number of parts to be assembled during one cycle; the larger the number of parts, the longer the cycle. Since, however, the number of operations at each station on a transfer system can be varied at the system design stage, this limitation does not exist. If necessary, for assemblies with large numbers of parts, transfer systems with large numbers of stations can be used to keep the cycle time low; this results in greater utilization of feeders.

REFERENCES

1. Boothroyd, G., Poli, C., and Murch, L.E., *Automatic Assembly*, Marcel Dekker, New York, 1982.
2. Murch, L.E., and Boothroyd, G., "On-Off Control of Parts Feeding," *Automation*, Vol. 18, Aug. 1970, pp. 32–34.
3. Redford, A.H., and Boothroyd, G., "Designing Gravity Feed Tracks for Headed Parts," *Automation*, Vol. 17, May 1970, pp. 96–101.
4. Arnstrom, A., and Grondahl, P., "Magazines for Flexible Automatic Assembly," IVF - Result 83607, Stockholm, 1983.
5. Suzuki, T., and Kohno, M., "The Flexible Parts Feeder Which Helps a Robot Assemble Automatically," *Assembly Automation*, Vol. 1, No. 2, 1981.
6. Redford, A.H., Lo, E.K., and Killeen, P., "Cost Analysis for Multi-Arm Robotic Assembly," *Assembly Automation*, Vol. 3, No. 4, p. 202.
7. Heginbotham, W.B., Barnes, D.F., Purdue, D.R., and Law, D.J., "Flexible Assembly Module with Vision Controlled and Placement Device," Production Engineering Research Association, Melton Mowbray, Leicestershire, England.
8. Zenger, D., and Dewhurst, P., "Automatic Handling of Parts for Robot Assembly," Annals of the C.I.R.P. Vol. 33, No. 1, 1984, p. 279.
9. Lennartz, C., and Boothroyd, G., "Economics of Part Presentation for Robot Assembly," Dept. of Mechanical Engineering, Univ. of Massachusetts, Amherst, Mass., 1984.
10. Boothroyd, G., "Economics of General-Purpose Assembly Robots," Annals of the C.I.R.P. Vol. 33, No. 1, 1984, p. 287.
11. Taniguchi, N., "Present State of the Arts of System Design on Automated Assembly in Japan," *Proceedings of the 4th International Conference on Assembly Automation*, Tokyo, Oct. 1983, p. 1.
12. Warnecke, H.J., and Walther, J., "Assembly of Car Engines by Hand in Hand Working Robots," *Proceedings of the 4th International Conference on Assembly Automation*, Tokyo, Oct. 1983, p. 15.

6

Performance and Economics
of Assembly Systems

Multistation assembly machines may be classified into two main groups according to the system used to transfer assemblies from workstation to workstation. The first group includes those assembly machines that transfer all the work carriers simultaneously. These are known as indexing or synchronous machines, and a stoppage of any individual workhead causes the whole machine to stop. In the other group of machines, which are known as free-transfer or nonsynchronous assembly machines, the workheads are separated by buffers containing assemblies, and transfer to and from these buffers occurs when the particular workhead has completed its cycle of operations. Thus, with a free-transfer machine, a fault or stoppage of a workhead will not necessarily prevent another workhead from operating because a limited supply of assemblies will usually be available in the adjacent buffers.

One of the principal problems in applying automation to the assembly process is the loss in production resulting from stoppages of automatic workheads when defective component parts are fed to the machine. With manual workstations on an assembly line, the assembly workers are able to discard defective parts quickly, with little loss of production. However, a defective part fed to an automatic workhead can, on an indexing machine, cause a stoppage of the whole machine, and production will cease until the fault is cleared. The resulting downtime can be very high with assembly machines having several automatic workheads. This can result in a serious loss in production and a consequent increase in the cost of assembly. The quality levels of the parts to be used in automatic assembly must, therefore, be considered when an assembly machine is designed.

In the following sections, a study is made of the effects of the quality levels of parts on the performance and economics of assembly machines.

6.1 INDEXING MACHINES

6.1.1 Effect of Parts Quality on Downtime

In the following analysis, it will be assumed that an indexing machine having n automatic workheads and operating with a cycle time of t sec is fed with parts having, on average, a ratio of defective parts to acceptable parts of x. It will also be assumed that a proportion m of the defective parts will cause machine stoppages and, further, that it will take an operator T sec, on average, to locate the failure, remove the defective part, and restart the machine.

With these assumptions, the total downtime due to stoppages in producing N assemblies will be given by $mxnNT$. Each time the machine indexes, all assembly tasks are completed and one assembly is delivered from the machine; hence, the machine time to assemble N assemblies is Nt sec and, thus, the proportion of downtime D on the machine is given by

$$D = \frac{\text{Downtime}}{\text{Assembly time} + \text{downtime}}$$

$$= \frac{mxnNT}{Nt + mxnNT} = \frac{mxn}{mxn + t/T} \tag{6.1}$$

In practice, a reasonable value of the machine cycle time t would be 6 sec, and experience shows that a typical value for the average time taken to clear a fault is 30 sec. With these figures, the ratio t/T will be 0.2, and Fig. 6.1 shows the effect of variations in the mean quality level of the parts on the downtime for indexing machines with 5, 10, 15, and 20 automatic workheads. (It is assumed in this example that all defective parts will produce a stoppage of the machine and, thus, $m = 1$.)

For standard fasteners such as screws, which are often employed in assembly processes, an average value for x might be between 0.01 and 0.02. In other words, for every 100 acceptable screws, there would be between one and two defective ones. A higher quality level is generally available, but with screws, for example, a reduction of x to 0.005 may double their price and seriously affect the cost of the final assembly. It will be seen later that a typical economic value for x is 0.01, and Fig. 6.1 shows that, with this value, the

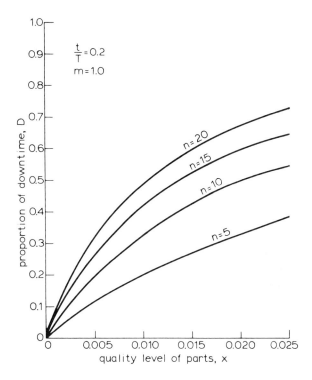

Fig. 6.1 Effect of parts quality on indexing machine downtime; n is the number of automatic workheads, t the machine cycle time, T the time to correct a machine fault, and m the proportion of defective parts causing a machine stoppage.

downtime on an assembly machine having 10 automatic workheads is 0.33 of the total time available. These results show why it is rarely economical to use indexing machines having a large number of automatic workheads. They also illustrate why, in practice, it is advisable to allow for downtime when considering the use of an indexing assembly machine.

6.1.2 Effects of Parts Quality on Production Time

In the example above, it was assumed that all the defective parts fed to the automatic workheads would stop the machine [that is, $m = 1$ in Eq. (6.1)]. In practice, however, some of these defective parts would pass through the feeding devices and automatic workheads but would not be assembled correctly and would result in the production of an unacceptable assembly. In this case, the effect of the defective part would be to cause downtime on the machine equal to only one machine cycle. The time taken to produce N assemblies, whether these are acceptable or not, is given by $(Nt + mxnNT)$ and, if $m < 1$, only about $[N - (1 - m)xnN]$ of the assemblies produced will be acceptable. The average production time t_{pr} of acceptable assemblies is therefore given by

$$
\begin{aligned}
t_{pr} &= \frac{Nt + mxnNT}{N - (1 - m)xnN} \\
&= \frac{t + mxnT}{1 - (1 - m)xn}
\end{aligned}
\tag{6.2}
$$

Taking typical values of $x = 0.01$, $t = 6$ sec, $T = 30$ sec, and $n = 10$, Eq. (6.2) becomes

$$
t_{pr} = \frac{30(2 + m)}{9 + m}
\tag{6.3}
$$

Equation (6.3) is plotted in Fig. 6.2 to show the effect of m on t_{pr}, and it can be seen that, for a maximum production rate of acceptable assemblies, m should be as small as possible. In other words, when designing the workheads for an indexing assembly machine when a high production rate is required, it is preferable to allow a defective part to pass through the feeder and workhead and "spoil" the assembly rather than allow it to stop the machine. However, in practical circumstances, the cost of dealing with the unacceptable assemblies

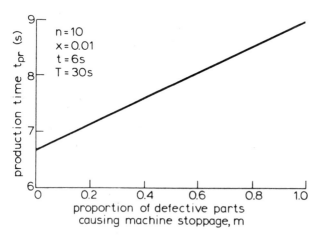

Fig. 6.2 Variation in production time t_{pr} with changes in the proportion of defective parts causing a machine fault; n is the number of automatic workheads, x the ratio of defective to acceptable parts, t the machine cycle time, and T the time to correct a machine fault.

produced by the machine must be taken into account, and this will be considered later.

For the case in which the defective parts always stop the machine, m is equal to 1, and Eq. (6.2) becomes

$$t_{pr} = t + xnT \qquad (6.4)$$

Figure 6.3 now shows how the production rate ($1/t_{pr}$) is affected by changes in nominal machine rate ($1/t$) for various values of x and for typical values of $T = 30$ sec and $n = 10$. It can be seen that, when x is small, the production rate approaches the machine speed. However, in general, x will lie within the range 0.005–0.02 and, under these circumstances, it can be seen that, for high machine speeds (short cycle times), the production rate tends to become constant. Alternatively, it may be stated that, as the cycle time is reduced for otherwise constant conditions, the proportion of downtime increases, and this results in a relatively small increase in the production rate. This explains why it is rarely practicable to have indexing assembly machines working at very high speeds.

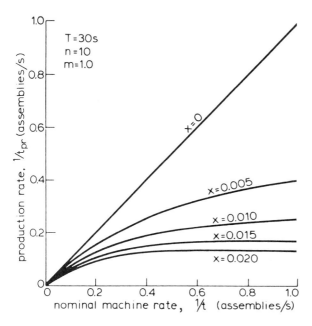

Fig. 6.3 Effect of parts quality level on indexing machine production rate; n is the number of automatic workheads, T the time to correct a machine fault, and m the proportion of defective parts causing a machine stoppage.

6.1.3 Effect of Parts Quality on the Cost of Assembly

The total cost C_t of each acceptable assembly produced on an assembly machine (where each workhead assembles one part) is given by the sum of the costs of the individual parts $C_1 + C_2 + C_3 + \cdots + C_n$ plus the cost of operating the machine for the average time taken to produce one acceptable assembly. Thus,

$$C_t = M_t t_{pr} + C_1 + C_2 + C_3 + \cdots + C_n \qquad (6.5)$$

where M_t is the total cost of operating the machine per unit time and includes operators' wages, overhead charges, actual operating costs, machine depreciation, and the cost of dealing with the unacceptable assemblies produced, and t_{pr} is the average production time of acceptable assemblies and may be obtained from Eq. (6.2).

In estimating M_t, it will be assumed that a machine stoppage caused by a defective part will be cleared by one of the operators employed on the machine and that no extra cost will be entailed other than that due to machine downtime. Further, it will be assumed that if a defective part passes through the workhead and spoils an assembly, it will take an extra assembly worker t_c sec to dismantle the assembly and replace the nondefective parts back in the appropriate feeding devices.

Thus, the total operating cost M_t is given by

$$M_t = M + P_u t_c W_a \qquad (6.6)$$

where M is the cost of operating the machine per unit time if only acceptable assemblies are produced and W_a is the assembly worker's rate, including overhead. The number of unacceptable assemblies produced per unit time is denoted by P_u and is given by

$$P_u = \frac{(1 - m)xn}{t + mxnT} \qquad (6.7)$$

Substitution of Eq. (6.7) in Eq. (6.6) gives

$$M_t = M + \frac{(1 - m)xnt_c W_a}{t + mxnT} \qquad (6.8)$$

In estimating the cost C_i of an individual component part, it will be assumed that this can be broken down into (1) the basic cost of the part, irrespective of quality level, and (2) a cost that is inversely proportional to x and that will therefore increase for better-quality parts. Thus, the cost of each part may be expressed as

$$C_i = A_i + \frac{B}{x} \qquad (6.9)$$

In this equation, B is a measure of the cost due to quality level and, for the purposes of the present analysis, will be assumed constant, regardless of the basic cost A_i of the parts.

If Eqs. (6.2), (6.8), and (6.9) are now substituted into Eq. (6.5), the total cost C_t of each acceptable assembly becomes, after rearrangement,

$$C_t = \frac{M(t + mxnT) + (1 - m)xnt_c W_a}{1 - (1 - m)xn} + \sum_{i=1}^{n} A_i + \frac{nB}{x} \qquad (6.10)$$

Equation (6.10) shows that the total cost of an assembly can be broken down as follows:

1. A cost that will decrease as x is reduced
2. A cost that is constant
3. A cost that will increase as x is reduced

It follows that, for a given situation, an optimum value of x will exist that will give a minimum cost of assembly. For the moment, the optimum value of x will be considered for the case in which $m = 1$ (that is, where all defective parts cause a stoppage of the machine).

With $m = 1$, Eq. (6.10) becomes

$$C_t = Mt + Mxnt + \frac{nB}{x} + \sum_{i=1}^{n} A_i \qquad (6.11)$$

Cost of	+	Cost of	+	Cost of	+	Basic
assembly		downtime		parts		cost of
operations				quality		parts

Equation (6.11) is now differentiated with respect to x and set equal to zero, yielding the following expression for the optimum value of x, giving the minimum cost of assembly:

$$x_{opt} = \left[\frac{B}{MT} \right]^{1/2} \qquad (6.12)$$

It is interesting to note that, for a given assembly machine, where M and B are constants, the optimum quality level of the parts used is dependent only on the time taken to clear a defective part from a workhead.

Figure 6.4 shows how the cost of a part might increase as the quality level is improved. In this case, $A_i = \$0.002$ and $B = \$0.00003$. If typical values of $M = 0.01$ \$/sec (36 \$/hr) and $T = 30$ sec are now substituted in Eq. (6.12), the corresponding optimum value of x is 0.01 (1%). Equation (6.12) may now be substituted in Eq. (6.11) to give an expression for the minimum cost of assembly:

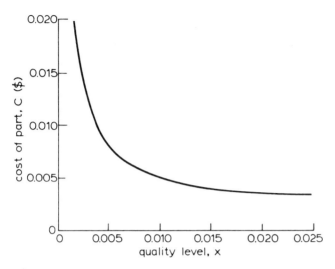

Fig. 6.4 Typical relationship between part quality and cost.

$$C_t(\min) = Mt + 2n(MBT)^{1/2} + \sum_{i=1}^{n} A_i \qquad (6.13)$$

With $t = 6$ sec and $n = 10$, the cost of assembling each complete set of parts [the first two terms in Eq. (6.13)] would be \$0.12. Half of this cost would be attributable to the assembly operation itself, and the other half to the increased cost due to parts quality and the corresponding cost of machine downtime. Figure 6.5, where the first three terms in Eq. (6.11) are plotted, shows how these individual costs would vary as the quality level x varies, using the numerical values quoted in the example above. In this case, if parts having 0.02 defective were to be used instead of the optimum value of 0.01, the cost of assembly would increase by approximately 12%. This is a variation of \$0.015 per assembly and, with an average production time of 9 sec [calculated from Eq. (6.2)], it represents an extra expense of approximately 12,000k\$/yr per shift.

In the analysis above, it was assumed that all defective parts would stop the machine. If, instead, it were possible to allow these parts to pass through the automatic devices and spoil the assemblies, the cost of assembly could be obtained by substituting $m = 0$ into Eq. (6.10). Thus,

$$C_t = \frac{Mt + xnt_cW_a}{1 - xn} + \sum_{i=1}^{n} A_i + \frac{nB}{x} \qquad (6.14)$$

Again, an optimum value of x arises and is found by differentiation of Eq. (6.14) to be

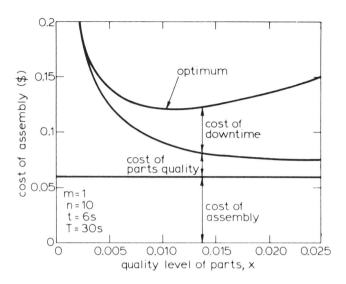

Fig. 6.5 Effect of parts quality level on assembly costs.

$$x_{opt} = \left[n + \left(\frac{t_c W_a}{B} + \frac{Mt}{B} \right)^{1/2} \right]^{-1} \tag{6.15}$$

Taking $x = 0.002$ \$/sec, $t_c = 60$ sec, and the remaining figures as before, x_{opt} is found to be approximately 0.011. From Eq. (6.14), the minimum cost of assembly is \$0.109, which represents a saving of 9% on the cost of assembly compared to the preceding example.

It is possible to draw two main conclusions from the examples discussed. First, for the situation analyzed, it would be preferable to allow a defective part to pass through the workhead and spoil the assembly rather than to allow it to stop the indexing machine. This would not only increase the production rate of acceptable assemblies but would also reduce the cost of assembly. Second, an optimum quality level of parts always exists that will give minimum cost of assembly.

6.2 FREE-TRANSFER MACHINES

A free-transfer or nonsynchronous machine always gives a higher production rate than the equivalent indexing machine. This is because the buffers of assemblies available between workheads will, for a limited time, allow the continued operation of the remaining workheads when one has stopped. Provided that the buffers are sufficiently large, the stopped workhead can be restarted before the other workheads are affected and the downtime on the machine approaches the downtime on the workhead that has the most stoppages. The following analysis will show that, even with relatively small buffers, the production rate of the machine can be considerably higher than that obtained with the equivalent indexing machine.

It will be assumed in the analysis that all the workheads on a free-transfer machine are working at the same cycle time of t seconds. Each workhead is fed with parts having the same quality level of x (where x is the ratio of defective to acceptable parts), and between each pair of workstations is a buffer that is large enough to store b assemblies. Any workhead on a free-transfer assembly machine will be forced to stop under three different circumstances.

1. A defective part is fed to the workhead and prevents the completion of its cycle of operations. In this case it will be assumed that an interval of T seconds elapses before the fault is cleared and the workhead restarted.
2. The adjacent workhead up the line has stopped and the supply of assemblies in the buffer storage between is exhausted.
3. The adjacent workhead down the line has stopped, and the buffer between is full.

Consideration of these last two circumstances leads to the conclusion that spaces in buffers cause the same problems as assemblies in buffers. Thus, to optimize the performance of a free-transfer machine, the number of spaces in buffers should be made equal to the number of assemblies in buffers. In other words, the buffers should be half-filled with assemblies. It will also be assumed that, at any time, the buffer between any two workheads is half-filled. Of course, this will rarely be the case and, in fact, for half of the time, the buffer will be empty and, for the remaining time, it will be full. In the following analysis, the assumption that any particular buffer will be half-filled with assemblies will, therefore, lead to an underestimation of the downtime caused by workhead faults.

Thus, if two adjacent workheads have $b/2$ assemblies in the buffer storage between, then a fault in the first workhead will prevent the second from working after a time delay of $bt/2$ sec. A fault in the second workhead will prevent the first from working after the same time. Also, over a long period, the average downtime on each workhead must be the same. It will also be assumed that when a fault occurs, there will always be a technician available to correct it and that no workhead will stop while another is stopped. The errors resulting from this latter assumption will become large when the quality level of the parts is poor (large x) and with a large number of automatic workheads (large n). However, specimen calculations show that these errors are negligible with practical values of x and n and produce an overestimate of the machine downtime.

6.2.1 Performance of a Free-Transfer Machine

For a typical workhead on a free-transfer machine (Fig. 6.6) producing N assemblies, Nx stoppages will occur if m is unity. If each fault takes T sec to correct, the downtime on the first workhead due to its own stoppages is given by NxT. This same average downtime will apply for stoppages on the second workhead down the line, but the first will be prevented from working only for a period of $Nx[T - bt/2]$ sec. Similarly, stoppages of the third workhead will prevent the first from working only for a period of $Nx[T - bt]$ sec. The same expressions can be derived for the effects of the second and third workheads up the line. Thus, it can be seen that the total downtime d on any workhead while the machine produces N assemblies is given by

$$\frac{d}{Nx} = \cdots \left[T - \frac{3bt}{2} \right] + [T - bt] + \left[T - \frac{bt}{2} \right] + T$$
$$+ \left[T - \frac{bt}{2} \right] + [T - bt] + \left[T - \frac{3bt}{2} \right] + \cdots$$

or

$$\frac{d}{Nx} = T + [2T - bt] + [2T - 2bt] + [2T - 3bt] + \cdots \qquad (6.16)$$

It should be noted that if any term in square brackets is negative, it should be omitted. It is necessary at this stage to know the relative values of T and t. If, for example, the ratio T/t is 5 and $b = 4$, Eq. (6.16) becomes

$$\frac{d}{Nx} = 5T - 3bt \qquad (6.17)$$

or

$$d = 13Nxt \qquad (6.18)$$

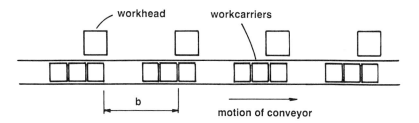

Fig. 6.6 Free-transfer machine showing the average situation during operation for $b = 4$.

The proportion of downtime D on the machine may now be obtained as follows:

$$D = \frac{\text{Downtime}}{\text{Assembly time} + \text{downtime}} = \frac{d}{Nt + d} \tag{6.19}$$

and substitution of d from Eq. (6.18) gives

$$D = \frac{13xT}{t + 13xT} \tag{6.20}$$

For different values of b, other bracketed terms in Eq. (6.16) will become zero or negative, and new values of d will be obtained.

In general, the machine downtime d may be expressed as

$$d = KNxt \tag{6.21}$$

where K is a factor that depends on the values of T/t and b. Table 6.1 gives the values of K for various values of b when $T/t = 5$.

The proportion of downtime D is now given by

$$D = \frac{d}{Nt + d} = \frac{Kx}{1 + Kx} \tag{6.22}$$

For $b \geqslant 10$, all bracketed terms are omitted from Eq. (6.16), and the equation for the downtime on each workhead becomes

$$\frac{d}{Nx} = T \tag{6.23}$$

The downtime on the machine equals the downtime on any workhead and, thus, for $b \geqslant 10$,

$$D = \frac{5x}{1 + 5x} \tag{6.24}$$

Figure 6.7 shows how the proportion of downtime is affected by the size of the buffers for $x = 0.01$ and for machines with 5, 11, 21, and 41 stations. It can

Table 6.1 Relationship Between Buffer Capacity b and the Factor K

b	K
0	25
2	19
4	13
6	9
8	7
10	5

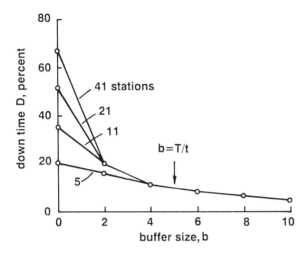

Fig. 6.7 Theoretical effect of buffer storage size on proportion of downtime for several free-transfer machines.

be seen that significant improvements in performance can be obtained with only small buffers, especially with large machines. Also, with large buffers, the machine downtime is independent of the machine size and approaches the downtime on a single station when it is unaffected by stoppages at other stations.

In theory, when the size of the buffer storage b is greater than or equal to $2T/t$, the workheads are completely isolated one from the other, and further increases in the size of b have no effect. However, the greatest benefit occurs with the smaller buffers. Therefore, as a practical guide, it is often assumed that, for a good free-transfer assembly machine design, b equals T/t.

It should be pointed out that the method of analysis for free-transfer machines presented in this chapter is only approximate although the results of computer simulations presented in Fig. 6.8 for the same conditions as those in Fig. 6.7 show that the analysis gives a good approximation to the true performance of a machine when $b \leqslant T/t$.

6.2.2 Average Production Time for a Free-Transfer Machine

When considering the economics of free-transfer machines, the cost of providing the buffers has to be taken into account. Ideally, it is necessary to develop a mathematical model in which the size b of the buffers between each station is a variable and then to study how the assembly costs C_{pr} are affected by the

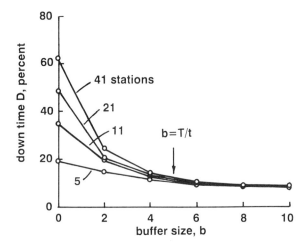

Fig. 6.8 Results of computer simulations [1] showing the effect of buffer storage on downtime for several free-transfer machines.

magnitude of b. However, later analysis will show that, when the number of parts to be assembled n is small, the free-transfer machine is uneconomical compared to an indexing machine. When n is large, the reverse is true. Thus, when n is small, the optimum value of b is zero (representing an indexing machine) and, when n is large, the optimum value is close to that at which the various workheads do not affect one another significantly. Inspection of Figs. 6.7 and 6.8, which show the effect of buffer size on the proportion of downtime, indicates that, when $b \geqslant 10$, the proportion of downtime D on the machine approaches that of a single isolated workhead. In general, this is true if $b \geqslant 2T/t$, where T is the downtime due to one defective part and t is the workhead cycle time. However, analysis of the economics of individual machines shows that a value of b equal to T/t results in cost savings closely approaching those for $b = 2T/t$ and, therefore, to simplify analysis and to make rough cost comparisons, it will be assumed that, on a free-transfer machine, $b = T/t$ and that the resulting downtime for the machine approaches twice that for an individual isolated station (Figs. 6.7 and 6.8).

Hence, for a free-transfer machine of any size, the average production time can be estimated from

$$t_{pr} = t + 2xT \tag{6.25}$$

where x is the parts quality level. It is, of course, assumed here that all defective parts will cause a stoppage of the workhead. Thus, Eq. (6.25) can be compared with Eq. (6.4) for an indexing machine.

6.2.3 Number of Personnel Needed for Fault Correction

In the computer simulations, the assembly machines were provided with an excess of personnel so that all faults caused by defective parts would receive immediate attention. However, the number of personnel needed in order to achieve this situation on large systems would be uneconomical, and it would be better to allow some faults to go unattended for a short time in order to increase the utilization of the fewer personnel. Clearly, for optimum working, the number of personnel required will be proportional to the number of stations on the machine.

During the production of N assemblies, each containing n parts automatically assembled, and with each part having x defectives causing faults, the total faults will be given by Nxn, and the total time spent by personnel correcting faults will be $NxnT$, where T is the average time to correct each fault. Since the total production time on a free-transfer machine is given by Eq. (6.25), the minimum number of technicians N_{tech} required to correct faults is given by

$$N_{\text{tech}} = \frac{\text{Total fault correction time}}{\text{Total production time}}$$

$$= \frac{NxnT}{N(2xT + t)}$$

$$= \frac{xn}{2x + t/T} \tag{6.26}$$

Alternatively, the maximum number of stations $n_{s\,\text{max}}$ that one technician can tend is given by

$$n_{s\,\text{max}} = \frac{n}{N_{\text{tech}}} = 2 + \frac{t}{xT} \tag{6.27}$$

Using $t/T = 0.2$ and $x = 0.01$, Eq. (6.27) would show that one technician could tend no more than 22 stations on a free-transfer machine having these characteristics. However, since faults can occur on a free-transfer machine while an earlier fault is still being corrected, the number of stations per technician will be less than that given by Eq. (6.27). Figure 6.9 shows that the optimum number of stations is closer to 16. Thus, Eq. (6.27) could be modified to give the economical number of stations per technician as follows:

$$n_s = 2 + \frac{2t}{3xT} \quad \text{approximately} \tag{6.28}$$

With an indexing machine, any fault causes the whole machine to stop and, therefore, no more than one technician, theoretically, would be needed to correct faults.

We shall now use these results to compare the economics of various assembly systems.

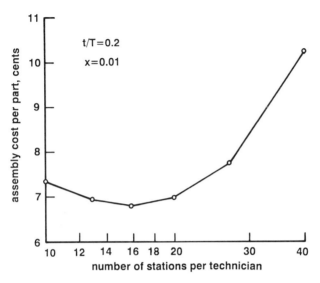

Fig. 6.9 Effect of number of technicians available to clear faults on a free-transfer machine (from computer simulation).

6.3 BASIS FOR ECONOMIC COMPARISONS OF AUTOMATION EQUIPMENT

In simple analyses of the economics of automation equipment, it is advantageous to convert the capital cost of the equipment to an equivalent assembly worker rate. For this purpose, a factor Q_e can be defined as the cost of the capital equipment that can economically be used to do the work of one assembly worker on one shift [2,3]. This figure, which must be determined for a particular company, is the basis for the following economic comparisons. For example, if a simple piece of automation equipment is being considered and it can do the job of one assembly worker, then the economical cost of the equipment would be Q_e for one shift working, $2Q_e$ for two shifts working, etc. The rate for a piece of equipment initially costing C_e is then given by $C_e W_a / (S_n Q_e)$, where W_a is the rate for one assembly worker and S_n is the number of shifts. In fact, many companies will determine Q_e by multiplying the annual cost of an assembly worker by a factor that usually lies between 1 and 3, representing a payback period of 1–3 yr.

6.3.1 Basic Cost Equations

The cost of assembly C_a for a complete assembly is given by

$$C_a = t_{\text{pr}} \left[W_t + \frac{C_e W_a}{S_n Q_e} \right] \tag{6.29}$$

where t_{pr} = the average time between delivery of complete assemblies for a fully utilized system

W_t = the total rate for the machine personnel

C_e = the total capital cost for all equipment, including engineering setup and debugging cost

For the purpose of comparing the economics of assembly systems, the cost of assembly per part will be used and will be nondimensionalized by dividing this cost per part by the rate for one assembly worker W_a and the average manual assembly time per part t_a. Thus, the dimensionless assembly cost per part C_d is given by

$$C_d = \frac{C_a}{n W_a t_a} \qquad (6.30)$$

Substitution of Eq. (6.29) into Eq. (6.30) gives

$$C_d = \left[\frac{t_{pr}}{t_a} \right] \left[W_r + \frac{C_e}{n S_n Q_e} \right] \qquad (6.31)$$

where

$$W_r = \frac{W_t}{W_a n} \qquad (6.32)$$

and is the ratio of the cost of all personnel compared with the cost of one manual assembly worker and expressed per part in the assembly. Thus, the dimensionless assembly cost per part for an assembly worker working without any equipment will be unity, which forms a useful basis for comparison purposes.

Finally, it should be pointed out that, for a particular assembly machine, Eq. (6.31) holds true only if the required average production time t_q for one assembly is greater than or equal to the minimum production time t_{pr} obtainable for the machine. In other words, if $t_{pr} \leq t_q$, then t_q must be substituted for t_{pr} in Eq. (6.31) because the machine is not fully utilized. If $t_{pr} > t_q$, then more than one machine will be needed to meet the required production rate. Now, for example, two machines producing N assemblies/hr will give the same assembly costs per assembly as one machine producing $N/2$ assemblies/ hr. Hence, it will be assumed that, as the average required production time for one assembly t_q is reduced below the production time t_{pr} obtainable from one machine, the assembly costs become constant and are given by Eq. (6.31). To obtain t_q, it is necessary to know the required annual production volume per shift V_s and the plant efficiency P_e, the latter being defined as the time actually worked in the plant divided by the time available. Thus,

$$t_q = \frac{0.072 P_e}{V_s} \qquad (6.33)$$

In this formula, t_q is given in seconds, V_s is in millions, P_e is expressed as a percentage, and the factor 0.072 arises from the assumption that 7.2 Msec is the maximum time available in one shift-year.

We have now established the basic cost equations necessary to compare the economics of assembly systems using the capital equivalent of assembly worker costs.

6.4 COMPARISON OF INDEXING AND FREE-TRANSFER MACHINES

In analyzing the economics of special-purpose assembly machines, the following nomenclature will be used, and the numerical values given in parentheses represent the estimates of the cost of the equipment used in the later comparisons:

C_T = cost of transfer device per workstation for an indexing machine (10 k$)

C_B = cost of transfer device per space (workstation or buffer space) for a free-transfer machine (5 k$)

C_C = cost of work carrier (1 k$)

C_F = cost of automatic feeding device and delivery track (7 k$)

C_W = cost of workhead (10 k$)

W_{tech} = rate for one technician engaged in correcting faults on the machine (0.012 $/sec)

W_a = rate for one assembly worker (0.008 $/sec)

Q_e = equivalent cost of one assembly worker in terms of capital investment (90 k$)

t_a = average manual assembly time per part (8 sec)

It should be noted that equipment costs include the purchase costs, the assembly machine design, engineering and debugging costs, the costs of controls, and so on. It has been estimated [4] that purchase cost often forms only 40% of the total cost of equipment.

6.4.1 Indexing Machine

For an indexing machine, the total equipment cost C_e will be given by

$$C_e = n(C_T + C_W + C_F + C_C) \tag{6.34}$$

Assuming two assembly workers to load and unload assemblies, the total rate for the personnel will be

$$W_t = 2W_a + W_{tech} \tag{6.35}$$

Substituting for C_e from Eq. (6.34), for W_t from Eq. (6.35), and for t_{pr} from Eq. (6.4) into Eq. (6.31) gives, for the dimensionless cost of assembly per part,

$$C_d = \frac{(t + xnT)\left[2 + \dfrac{W_{tech}}{W_a} + \dfrac{n(C_T + C_W + C_F + C_C)}{S_n Q_e}\right]}{nt_a} \tag{6.36}$$

6.4.2 Free-Transfer Machine

For a free-transfer machine, where the number of work carriers chosen is such that, on average, the machine will be half-full, the total equipment cost C_e will be given by

$$C_e = n\left[C_W + C_F + (b + 1)\left(C_B + \frac{C_C}{2}\right)\right] \tag{6.37}$$

Again, assuming two assembly workers to load and unload assemblies and using Eq. (6.28), the total rate for the personnel will be

$$W_t = 2W_a + \left(2 + \frac{2t}{3xT}\right)W_{tech} \tag{6.38}$$

Substituting for $b = T/t$, t_{pr} from Eq. (6.25), C_e and W_t from Eqs. (6.37) and (6.38) into Eq. (6.31) gives

$$C_d = (t + 2xT)\left\{2 + \left[\frac{nW_{tech}}{2 + (2t/3xT)}\right]\left[\frac{1}{W_a}\right]\right.$$
$$\left. + n\,\frac{C_W + C_F + (1 + T/t)(C_B + C_C/2)}{S_n Q_e}\right\}nt_a \tag{6.39}$$

provided the number of technicians is not less than one. Assuming a cycle time t of 6 sec, a downtime T of 30 sec, a quality level x of 0.01, two shifts working ($S_n = 2$), and the numerical values of equipment and operator costs described earlier, Fig. 6.10 shows, for the two machines, how the dimensionless cost of assembly per part varies with the size of the machine. It can be seen that for small values of n (small machines), the indexing machine is the more economical of the two whereas, for large values of n, the free-transfer machine is the more economical. The rise in costs for the indexing machine is attributable entirely to the increasing downtime as the machine becomes larger. In fact, the region of the curve shown dashed represents downtime proportions of greater than 50%.

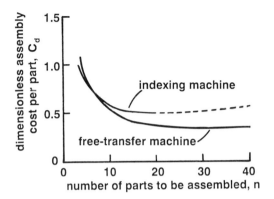

Fig. 6.10 Comparison of the economics of multistation assembly machines.

6.4.3 Effect of Production Volume

Turning now to the effects of production volume, it should first be realized that a special-purpose assembly machine is designed to work at a particular rate. This means that, under normal circumstances, the output can be obtained from a knowledge of t_{pr}, the mean production time obtainable. Thus, the obtainable annual production volume per shift V_s in millions is given by

$$V_s = 0.072 P_e / t_{pr} \tag{6.40}$$

In this formula, P_e is the plant efficiency (percent) and is defined as the time actually worked in the plant divided by the time available, and it is assumed that 7.2 million sec are available in one shift-year. Thus, on a graph of assembly cost versus annual production volume such as that shown in Fig. 6.11, a special-purpose assembly machine is represented by a single point corresponding to the production volume obtainable from the machine working at maximum capacity. However, for the purposes of comparison, it is desirable to be able to show the relation between assembly cost and the annual production volume per shift required.

If the volume required is less than that obtainable from the machine, then, as explained earlier, the assembly cost will be a function of the average production time required t_q, given by Eq. (6.33). In this case, when the assembly costs [given by Eq. (6.31)] and the annual production volume per shift are plotted on logarithmic scales, a linear relationship results as shown in Fig. 6.11. This relationship simply shows how costs increase because the machinery is not being fully utilized.

If the required production volume is greater than that obtainable from the machine, then it is assumed that backup will be provided in the form of

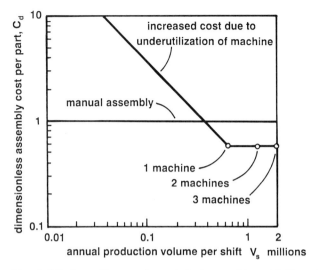

Fig. 6.11 Assembly costs for an indexing machine with 10 stations.

manual assembly stations. Then, when the required volume approaches twice that obtainable, a second machine will be employed, and so on. Since the assembly cost is independent of the number of assembly machines, the relation between dimensionless assembly cost and volume will be a horizontal line (Fig. 6.11) because of the smoothing provided by the use of manual assembly where required.

The multistation assembly machines considered above were both assumed to be "dedicated" to one product. However, it should be realized that the output from one machine is on the order of 1–5 million assemblies per year and that application of these systems is limited to mass production situations in which the product design is stable for a few years. It should also be realized that mass production constitutes only about 15–20% of all production in the United States and, for this reason, much attention is being given at present to the possibilities of automating batch assembly processes. It is clear that, for these applications, it will be necessary to employ programmable or flexible automation—in fact, assembly robots.

6.5 ECONOMICS OF ROBOT ASSEMBLY

One of the problems in making general economic comparisons between assembly systems is that, typically, some of the operations required for a particular product cannot be automated; one of the key characteristics of a system is its ability to handle this situation. Accordingly, a study was made [5] in which, by working closely with various key manufacturing organizations, the author was able to determine the kinds of assemblies that were generally considered candidates for robot assembly. It was found that these assemblies have similar characteristics and that a typical profile could be deduced for a candidate assembly. Using this typical candidate assembly, the economics of some configurations of robot assembly systems were considered. The following discussion will be restricted to those situations in which general-purpose assembly robots can be applied, and three basic systems will be modeled. These systems are:

1. A single-station machine with one fixture and one robot arm (Fig. 6.12)
2. A single-station machine with one fixture and two robot arms (Fig. 6.13)
3. A multistation free-transfer machine with single robot arms at the various stations where appropriate (Fig. 6.14)

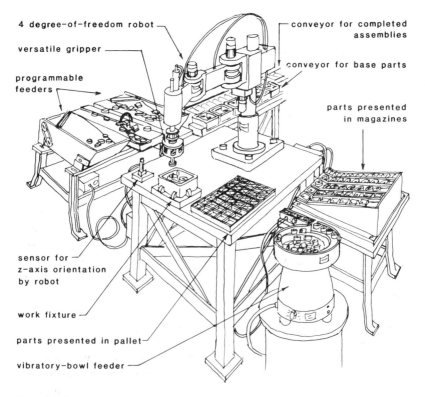

Fig. 6.12 Single-station assembly system with one robot arm.

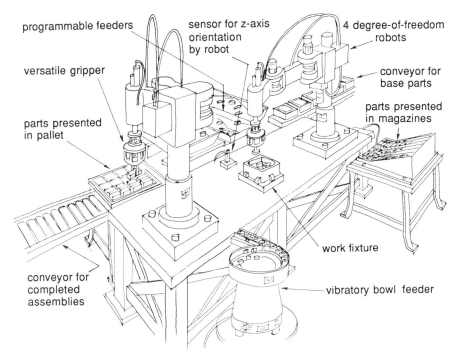

programmable feeders

sensor for z-axis
orientation
by robot

4 degree-of-freedom
robots

versatile gripper

conveyor for
base parts

parts presented
in magazines

parts presented
in pallet

conveyor for
completed
assemblies

work fixture

vibratory bowl feeder

Fig. 6.13 Single-station assembly system with two robot arms.

6.5.1 Parts Presentation

When the suitability of products for robot assembly is determined, it is important to consider the available methods of parts presentation. These range from "bin-picking," using the most sophisticated vision systems, to the manual loading of parts into pallets or trays.

Much research has been conducted on the subject of programmable feeders, and the impression is given that these can be used for systems that can assemble a variety of products. However, investigations have shown that most, if not all, of the "programmable" assembly systems presently being developed in the United States will be devoted to the assembly of one product or one family of products. Thus, these systems will be flexible in the sense that they can be adjusted quickly to accommodate different members of the product family (different styles) or to accommodate product design changes. Nevertheless, the equipment used on these systems will be "dedicated" to the product family, regardless of whether it is "special-purpose" equipment or "programmable" equipment.

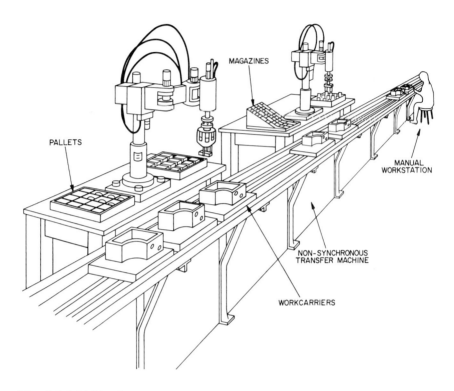

Fig. 6.14 Multistation free-transfer system utilizing robot arms where appropriate.

When the economic choice of parts-presentation method is to be determined, there will usually be only two basic types to consider:

1. Those involving manual handling and loading of individual parts into part trays, pallets, or magazines
2. Those involving automatic feeding and orienting of parts from bulk

In some cases, it is possible to obtain parts premagazined. Since these magazines have usually been filled by hand or automatically from bulk, then the cost of the magazining is borne by the supplier and is included in the cost of the parts. Thus, the economics of using premagazined parts will have to be considered on a case-by-case basis with a knowledge of the increased parts costs.

For the present purposes, no distinction will be made between different types of automatic equipment because the cost of feeding each part automatically, C_f, will be given by multiplying the rate for the equipment (cents/sec) by the time interval between which parts are required. Hence,

$$C_f = \frac{C_F W_a}{S_n Q_e} t_{at} \qquad (6.41)$$

where C_f is the cost of feeding one part (cents); C_F the cost of the feeder for one part type, including tooling, engineering, debugging (k$); and t_{at} the average station cycle time (sec).

The cost C_{mm} of using a manually loaded magazine to present one part will be given by

$$C_{mm} = (C_M W_a / S_n Q_e) t_{at} + W_a t_m \qquad (6.42)$$

where C_M is the cost of the magazines (k$), t_m the manual handling and insertion time for loading one part into the magazines (sec), and W_a the assembly worker rate (cents/sec).

For a simple part, the design for assembly handbook [6] gives a value of t_m of approximately 2 sec. A typical value of W_a is 0.8 cents/sec.

It will be assumed that a vibratory-bowl feeder, when supplied with simple tooling, engineered, debugged, and fitted with the necessary delivery track, costs $7000 and that the cost of a special-purpose magazine or feed track is $1000. If, as before, $S_n = 2$ and $Q_e = 90$, the comparison of part-presentation costs shown in Fig. 6.15 is obtained.

It can be seen that the special-purpose feeder would be more economical than the manual loading of magazines for station cycle times less than about 60 sec. Also, for longer cycle times, the use of feeding equipment rapidly becomes exorbitant.

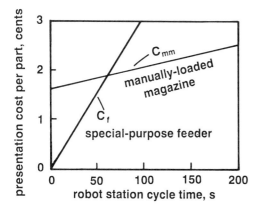

Fig. 6.15 Comparison of part-presentation costs for robot assembly.

6.5.2 Profile of Typical Candidate Assembly

In the study mentioned earlier, [5] companies known to be interested in robotic assembly were requested to submit examples of assemblies they considered possible candidates for robotic assembly. Analysis of 12 assemblies indicated that if n were the total number of parts, then:

1. $0.78n$ were different part types. This would approximate the case, for example, in which, in an assembly containing ten parts, three were identical. Information such as this is important because, for example, in a single-station machine using one arm, these three parts could all be presented by the same feeder.

2. $0.22n$ parts could be assembled automatically but not fed automatically and required manual handling. Such parts would be manually loaded into pallets, magazines, or feed tracks.

3. $0.71n$ parts could be fed and oriented using $0.5n$ special-purpose feeders.

4. $0.12n$ parts required manual handling and insertion or some manual manipulation. On a transfer machine, these would require a separate manual station. On a single-station system, the partially completed assemblies would be delivered to a manual station external to the system and then returned to the system after completion of the manual operation. Thus, two pick-and-place operations involving a robot arm would be required in addition to the manual operation.

5. $0.18n$ parts required special grippers or tools in addition to the basic grippers.

6. $0.15n$ parts required that the assembly be transferred to a special-purpose workhead or tool to complete the required operation.

7. The average assembly time per part for a single-station assembly system with one fixture and two robot arms was 3.29 sec; for a system with one arm, it was 5.4 sec.

8. If the product were to be manually assembled, the average assembly time per part would be 8.15 sec.

Using these figures, it is possible to estimate the cost of assembly using different configurations of robot assembly systems and to compare the results with the cost of manual assembly.

6.5.3 Single-Station Systems

Equipment Costs

First, considering a single-station system with one work fixture and one arm, the costs of the assembly equipment for an assembly containing n parts are listed in Table 6.2

Table 6.2 Breakdown of Equipment Costs for a One-Arm Single-Station System to Assemble n Total Parts

Item	Number required	Cost per unit, k$	Total cost, k$
Robot arm, sensors, controls, etc.	1	100	100
Work fixture	1	5	5
Standard grippers	1	5	5
Special grippers or tools	$0.18n$	5	$0.9n$
Magazines, pallets, etc.	$0.22n$	1	$0.22n$
Special-purpose feeders	$0.5n$	5	$2.5n$
Manual stations and associated feed tracks	$0.12n$	5	$0.6n$
Special-purpose workheads	$0.15n$	10	$1.5n$

Summing these costs gives the total system cost,

$$C_{e1} = 110 + 5.72n \qquad (6.43)$$

Similarly, the cost of a single-station system with one work fixture and two robot arms is found to be

$$C_{e2} = 165 + 6.62n \qquad (6.44)$$

The first term has increased because of the additional robot arm and gripper. The second term has increased because, with a two-arm system, in order to keep the cycle time low, it was found best to share repeated operations between the two arms and to duplicate the parts feeders.

Personnel Costs

One of the major problems in modeling automatic assembly systems is how to estimate the cost associated with the technician responsible for tending the machine and clearing stoppages due to faulty parts. On a multistation high-speed assembly machine, faults can occur so frequently that at least one full-time technician will be required who can also maintain adequate supplies of parts in the various automatic feeders. Under these circumstances, assemblies are produced at high rates, and the cost of the technician forms a relatively small portion of the total cost of assembly. However, with a low-speed single-station machine using robot arms, the cost of a full-time technician to tend the machine can constitute more than 50% of the cost of assembly.

For the purposes of this analysis, estimates will be made of the actual manual time involved in producing each assembly. Where parts are loaded into magazines or pallets or must be manually inserted, the total time required will be estimated. It will be assumed that if this total manual time is less than the system cycle time, the assembly worker can be engaged in other tasks unrelated to the particular assembly machine being modeled. Similarly, the time necessary for tending the machine will be estimated, and it will be assumed that the technician can be fully occupied by tending several such machines.

It should be emphasized that these assumptions will give the most optimistic estimates of assembly costs, especially while robot assembly is in the development stage. It can be argued, however, that, unless machines are eventually designed that can run with a minimum of attention, they will simply not provide an economical alternative to manual assembly. In fact, if a single-station general-purpose assembly machine were to require one full-time technician then, given assistance in the form of fixtures, etc., that same individual could probably perform the assembly tasks at a rate similar to that of the robot.

Part Quality

If the average time taken to clear a fault and restart the system is T, the number of parts in the assembly is n, and the average proportion of defective parts that will cause a fault is x, then the average production time per assembly t_{pr} will be given by

$$t_{pr} = t_{at} + xnT \qquad (6.45)$$

where t_{at} is the station cycle time (neglecting the downtime due to defective parts).

Basic cost equation

Using the basic equation for the dimensionless assembly cost per part (Eq. 6.31), we get

$$C_d = \frac{t_{pr}}{nt_a} \left[\frac{W_t}{W_a} + \frac{C_e}{S_n Q_e} \right] \qquad (6.46)$$

where C_e = total cost of equipment in k$ (including debugging, etc.), given by Eqs. (6.43) and (6.44)

W_t = total personnel rate given by $1.363nW_a + xnTW_{tech}$

Substitution of all these values and the equipment costs determined earlier into Eq. (6.46) gives the following: for a one-arm system,

$$C_{d1} = 0.662 + 0.022n \qquad (6.47)$$

and, for a two-arm system,

$$C_{d2} = 0.647 + 0.016n \qquad (6.48)$$

These equations are plotted in Fig. 6.16, where it can be seen that the economics of one- and two-arm single-station systems are very similar. Therefore, the decision as to which type of system to use must be made on an individual basis. However, it can be seen that single-station automatic systems used for assemblies containing more than about 18 parts are unlikely to be economical for the conditions assumed. An average station cycle time for such an assembly would be approximately 97 sec for a single-arm system and 59 sec for a two-arm system.

6.5.4 Multistation Transfer Systems

Equipment Costs

When the demand for a product is greater than can be assembled on a single-station assembly machine, duplicates of such machines can be installed or, alternatively, a multistation free-transfer machine can be employed. Our typical candidate assembly profile can again be used to study the economics involved in the latter choice.

On a free-transfer machine, it is usually not possible to perform manual operations at any station where automatic operations are carried out. Similarly, those operations involving special-purpose workheads must be carried out at separate stations. From the information on the profile of the candidate assembly, the number of manual stations will be $0.12n$, and the number of stations with special-purpose workheads will be $0.15n$. The remaining parts $(0.7n)$

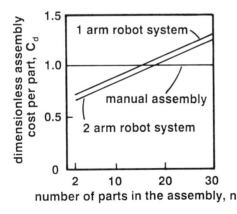

Fig. 6.16 Effect of number of parts on relative assembly costs for single-station robot systems.

will be assembled by robots, and the number of stations will depend on the required assembly rate. From Eq. (6.33), if the plant efficiency P_e is 80%, the station cycle time (neglecting downtime) would be $5.76/V_s$, where V_s is the annual volume per shift. The figure for the average assembly time per part for a one-arm system used earlier was 5.14 sec, so that the maximum number of parts assembled at one robot station n_r would be given by

$$n_r = \frac{5.76}{5.14V_s} = \frac{1.12}{V_s} \tag{6.49}$$

and the number of robot stations n_t on the machine would then be

$$n_t = \frac{0.7n}{n_r} = 0.625nV_s \tag{6.50}$$

Table 6.3 gives a breakdown of the equipment costs for a multistation free-transfer machine. Summing these equipment costs gives the total cost C_{et} for a free-transfer machine,

$$C_{et} = (12.77 + 78.5V_s)n \tag{6.51}$$

It should be noted that this equation is valid only for assemblies containing at least six parts and with station cycle time of at least 8.15 sec to allow for a typical manual assembly operation. This latter figure corresponds to an annual production volume per shift V_s of about 0.6 million.

Table 6.3 Breakdown of Equipment Costs for a Multistation Free-Transfer Machine to Assemble n Parts at a Rate of V_s Million Assemblies per Shift

Item	Number required	Cost per unit, k$	Total cost, k$
Robot arms, sensors, controls, etc.	$0.72nV_s$	75	$54nV_s$
Transfer device and 3 work carriers for each workstation	$0.72nV_s$ $+0.27n$	29	$20.9nV_s$ $+ 7.8n$
Grippers	$0.72nV_s$	5	$3.6nV_s$
Magazines, pallets, etc., for manual loading	$0.22n$	1	$0.22n$
Special-purpose feeders	$0.65n$	5	$3.25n$
Special-purpose workheads	$0.15n$	10	$1.5n$

Cost Equation

As with the analysis for single-station machines, it will be assumed that assembly workers are required only for the direct assembly and fault-correction procedures and can be engaged in tasks unrelated to the particular assembly for the remaining time. The downtime on a properly designed free-transfer machine approaches twice that for one individual station. Thus, the number of parts n in Eq. (6.45) should be regarded as twice the number of parts n assembled at one robot station and given by Eq. (6.49).

For a free-transfer machine, therefore,

$$t_{pr} = \frac{6.43}{V_s} \tag{6.52}$$

After substitution in Eq. (6.46), we get

$$C_{dt} = 0.567 + \frac{0.056}{V_s} \tag{6.53}$$

This result is plotted in Fig. 6.17 to show the effects of annual production volume on the cost of assembly. For comparison purposes, the results for single-station robot machines assembling 10 parts and for special-purpose machines assembling 10 parts are shown on the same figure.

It can be seen that the single-station robot machines and the special-purpose machines are each represented by one point on the graph. If these machines are underutilized, the cost of assembly increases rapidly. If volumes greater than those produced by one machine are needed, further identical machines could be used, giving the same assembly cost. However, with the multistation

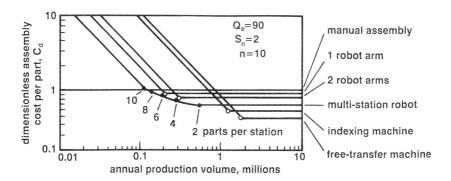

Fig. 6.17 Effect of annual production volume on relative cost of assembly.

robot machine, a degree of flexibility is available. By selecting the number of parts assembled at each station, the machine can be designed to match, within limits, the required production volume.

From Fig. 6.17, it can be seen that, for the conditions represented, special-purpose machines would be recommended for production volumes greater than about 1.5 million/yr. For volumes lower than about 200,000, manual bench assembly is preferable. Between these figures, assembly systems using robots might be economical.

REFERENCES

1. Boothroyd, G., and Dewhurst, P., "Design for Assembly Software Toolkit," Boothroyd Dewhurst, Wakefield, R.I., 1990.
2. Boothroyd, G., and Ho, C., "Performance and Economics of Programmable Assembly Systems" Society of Manufacturing Engineers, Technical Paper AD77–720, 1977.
3. Boothroyd, G., "Economics of Assembly Systems," *Journal of Manufacturing Systems*, Vol. 1, No. 1, 1982, pp. 111–127.
4. Lynch, P. M., "Economic Modeling and Design Criteria for Programmable Assembly Machines," Report T-625, The Charles Stark Draper Laboratory, Cambridge, Mass., June 1976.
5. Boothroyd, G., "Economics of General-Purpose Assembly Robots," *Annals of the CIRP*, Vol. 33, No. 1, 1984, p. 287.
6. Boothroyd, G., and Dewhurst, P., "Product Design for Assembly," Boothroyd Dewhurst, Wakefield, R.I., 1989.

7

Design for Manual Assembly

7.1 INTRODUCTION

Although there are many ways to increase manufacturing productivity (utilizing improved materials, tools, processes, plant layout, etc.), consideration of manufacturing and assembly during product design holds the greatest potential for significant reduction in production costs and increased productivity.

Robert W. Militzer, one-time president of the Society of Manufacturing Engineers, has stated [1]:

> ...as manufacturing engineers, we could do far more to improve productivity if we had greater input to the design of the product itself...it is the product designer who establishes the production [manufacture and assembly] process. For the fact is that much of the production process is implicit in product design.

In other words, if the product is poorly designed for manufacture and assembly, techniques can be applied only to reduce to a minimum the impact of the poor design. Improving the design itself is not worth considering at this late stage; usually, too much time and money have already been expended in justifying the design to consider major changes or even a completely new design. Only when manufacture and assembly techniques are incorporated early in the design stage (i.e., product design for ease of manufacture and assembly) will productivity be significantly affected.

7.2 WHERE DESIGN FOR ASSEMBLY FITS IN THE DESIGN PROCESS

The design process is an iterative, complex, decision-making engineering activity that leads to detailed drawings by which manufacturing can economically produce a quantity of identical products that can be sold. The design process usually starts with the identification of a need, proceeds through a sequence of activities to seek an optimal solution to the problem, and ends with a detailed description of the product.

The three main phases of the design process are the feasibility study, the preliminary design, and the detail design. Before any attempt is made to find possible solutions that may satisfy the hypothetical needs, the design problem is identified and formulated. During the feasibility study, plausible solutions are explored, and rough orders-of-magnitude checks are made on performance and perhaps product costs. Performance specifications for the most promising ideas are quantified during the preliminary design, and then the detail design results in the piece-part and assembly drawings that define all the characteristics of the product and its components.

Throughout the design process, the design team may need to factor many variables into their thinking and make numerous trade-offs with respect to performance, cost, reliability, service, and other attributes. Compared to these concerns, manufacturing and assembly costs tend to be difficult for design engineers to quantify and, as a result, often do not receive the attention they warrant.

Design for assembly (DFA) should be considered at all stages of the design process. As the design team conceptualizes alternative solutions and begins to realize their thoughts on paper, it should give serious consideration to the ease of assembly of the product or subassembly during production and during field service.

As concepts are analyzed against selected cost and performance criteria, a systematic analysis of product assemblability should be routinely performed. If cost or performance analysis requires that a concept be altered or redefined, then the efficiency of assembly of the reconceived design should be analyzed before final approval is made. As part of the detail design of parts and assemblies, part features, dimensions, and tolerances should be checked to make certain that they reflect the findings and conclusions of the DFA analysis.

Design engineers need a DFA tool to analyze effectively the ease of assembly of the products or subassemblies they design. The design tool should provide quick results and be simple and easy to use. It should ensure consistency and completeness in its evaluation of product assemblability. It should also eliminate subjective judgment from design assessment, allow free association of ideas, enable easy comparison of alternate designs, ensure that solutions are evaluated logically, identify assembly problem areas, and suggest alternate approaches for improving the manufacturing and assembly of the product.

By applying a DFA tool, communication between manufacturing and design engineering is improved, and ideas, reasoning, and decisions made during the design process become well documented for future reference.

The "Product Design for Assembly" handbook, [2] developed as a result of extensive research, provides methodological procedures for evaluating and improving product design for both economic manufacture and assembly. This goal is achieved by providing manufacturing input at the conceptualization stage of the design process in a logical and organized fashion. Another result of this approach is the availability of a clearly defined procedure for evaluating a design with respect to its ease of manufacture and assembly. In this manner, a feedback loop is provided to aid designers in measuring improvements resulting from specific design changes. This procedure also functions as a tool for motivating designers; through this approach, they can evaluate their own designs and, if possible, improve them. In both cases, the design is studied and improved at the conceptual stage and can be simply and inexpensively changed before involving manufacture and assembly. The "Product Design for Assembly" handbook attempts to meet these objectives by:

1.   Minimizing the dependence of the design engineer on the support of the manufacturing engineer by providing much of the assembly information needed to design new products for "ease of assembly."

2.   Guiding the designer to simplify the product so that savings in both assembly costs and piece-parts costs can be realized.

3.   Gathering information normally possessed by the experienced design engineer and arranging it, in a convenient way, for use by less experienced designers.

4.   Establishing a data base that consists of assembly times and cost factors for various design situations and production conditions.

The analysis of a product design for ease of assembly depends to a large extent on whether the product is to be assembled manually, with special-purpose automation, with general-purpose automation (robots), or with a combination of these. For example, the criteria for ease of automatic feeding and orienting are much more stringent than those for manual handling of parts. The following discussion will introduce design for manual assembly since it is always necessary to use manual assembly costs as a basis for comparison. In addition, even when automation is being seriously considered, some operations may have to be carried out manually, and it is necessary to include the cost of these operations in the analysis.

7.3 GENERAL DESIGN GUIDELINES FOR MANUAL ASSEMBLY

As a result of experience in applying DFA, it has been possible to develop general design guidelines that attempt to consolidate manufacturing knowledge and present them to the designer in the form of simple rules to be followed when creating a design. The process of manual assembly can be divided naturally into two separate areas: handling (acquiring, orienting, and moving the parts) and insertion and fastening (mating a part to another part or group of parts). The following design for manual assembly guidelines specifically address each of these areas.

A. Design Guidelines for Part Handling

In general, for ease of part handling, a designer should attempt to:

1. Design parts that have end-to-end symmetry and rotational symmetry about the axis of insertion. If this cannot be achieved, try to design parts having the maximum possible symmetry (see Fig. 7.1a).

2. Design parts that, in those instances in which the part cannot be made symmetric, are obviously asymmetric (see Fig. 7.1b).

3. Provide features that will prevent jamming of parts that tend to nest or stack when stored in bulk (see Fig. 7.1c).

4. Avoid features that will allow tangling of parts when stored in bulk (see Fig. 7.1d).

5. Avoid parts that stick together or are slippery, delicate, flexible, very small or very large, or that are hazardous to the handler (i.e., parts that are sharp, splinter easily, etc.) (see Fig. 7.2).

B. Design Guidelines for Insertion and Fastening

For ease of insertion, a designer should:

1. Design so that there is little or no resistance to insertion and provide chamfers to guide insertion of two mating parts. Generous clearance should be provided, but care must be taken to avoid clearances that will result in a tendency for parts to jam or hang up during insertion (see Figs. 7.3–7.6).

2. Standardize by using common parts, processes, and methods across all models and even across product lines to permit the use of higher-volume processes that normally result in lower product costs (see Fig. 7.7).

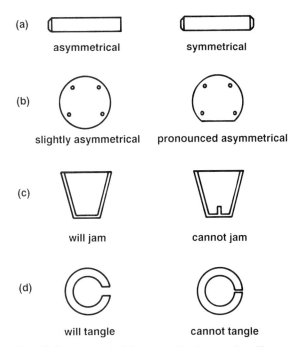

Fig. 7.1 Geometrical features affecting part handling.

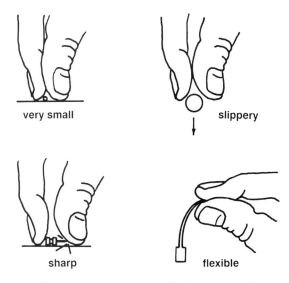

Fig. 7.2 Some other features affecting part handling.

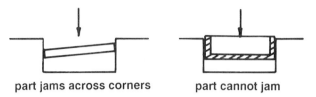

Fig. 7.3 Incorrect geometry can allow part to jam during insertion.

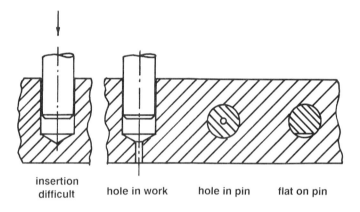

Fig. 7.4 Provision of air-relief passages to improve insertion into blind holes.

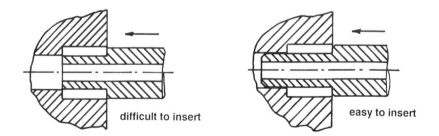

Fig. 7.5 Design for ease of insertion: assembly of long stepped bushing into counter-bored hole.

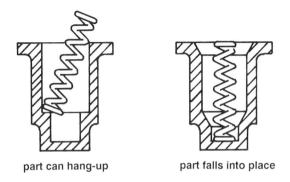

part can hang-up   part falls into place

Fig. 7.6 Provision of chamfers to allow easy insertion.

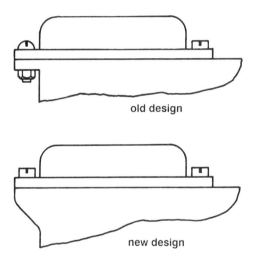

old design

new design

Fig. 7.7 Standardize parts.

3. Use pyramid assembly—provide for progressive assembly about one axis of reference. In general, it is best to assemble from above (see Fig. 7.8).

4. Avoid, where possible, the necessity for holding parts down to maintain their orientation during manipulation of the subassembly or during the placement of another part (see Fig. 7.9). If holding down is required, then try to design so that the part is secured as soon as possible after it has been inserted.

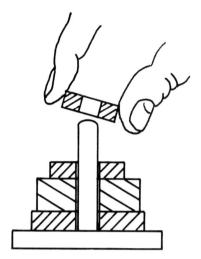

Fig. 7.8 Single-axis pyramid assembly.

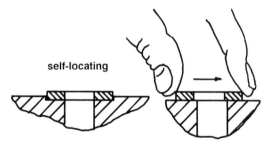

Fig. 7.9 Provision of self-locating features to avoid holding down and alignment.

5. Design so that a part is located before it is released. A potential source of problems in the placing of a part occurs when, because of design constraints, a part has to be released before it is positively located in the assembly. Under these circumstances, reliance is placed on the trajectory of the part being sufficiently repeatable to consistently locate it (see Fig. 7.10).

6. Consider, when common mechanical fasteners are used, the following sequence, which indicates the relative cost of different fastening

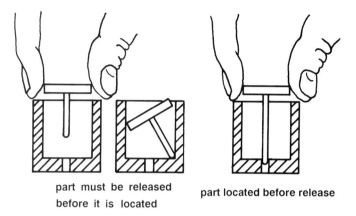

part must be released
before it is located

part located before release

Fig. 7.10 Design to aid insertion.

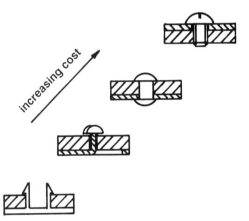

increasing cost

Fig. 7.11 Common fastening methods.

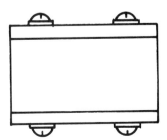

Fig. 7.12 Insertion from opposite directions requires repositioning of assembly.

processes, listed in order of increasing manual assembly cost (Fig. 7.11):

Snap fitting
Plastic bending
Riveting
Screwing

7. Avoid the need to reposition the partially completed assembly in the fixture (see Fig. 7.12).

Although functioning well as general rules to follow when design for assembly is carried out, guidelines are insufficient in themselves for a number of reasons. First, guidelines provide no means by which to evaluate a design quantitatively for its ease of assembly. Second, there is no relative ranking of all the guidelines that can be used by the designer to indicate which guidelines result in the greatest improvements in handling and assembly; there is no way to estimate the improvement resulting from the elimination of a part or from the redesign of a part for handling, etc. It is, then, impossible for the designer to know which guidelines to emphasize during the design of a product.

Finally, these guidelines are simply a set of rules that, when viewed as a whole, provide the designer with suitable background information to be used to develop a design that will be more easily assembled than a design developed without such a background. An approach must therefore be used that provides the designer with an organized method that not only encourages the design of a product that is easy to assemble but also provides an estimate of how much easier it is to assemble one design, with certain handling and assembly features, than to assemble another design with different handling and assembly features. The following describes the approach used in the "Product Design for Assembly" handbook that provides the means of quantifying assembly difficulty.

7.4 DEVELOPMENT OF THE SYSTEMATIC DFA METHODOLOGY

Starting in 1977, analytical methods were developed [3] for determining the most economical assembly process for a product and for analyzing ease of manual, automatic, and robot assembly. Experimental studies were performed [4–6] to measure the effects of symmetry, size, weight, thickness, and flexibility on manual handling time. Additional experiments were conducted [7] to quantify the effect of part thickness on the grasping and manipulation of a part using tweezers, the effects of spring geometry on the handling time of helical compression springs, and the effect of weight on handling time for parts requiring two hands for grasping and manipulation.

Regarding the design of parts for ease of manual insertion, experimental and theoretical analyses were performed [8–12] on the effect of chamfer design on manual insertion time, the design of parts to avoid jamming during assembly, the effect of part geometry on insertion time, and the effects of obstructed access and restricted vision on assembly operations.

A classification and coding system for manual handling, insertion, and fastening processes, based on the results of these studies, was presented in the form of a time standard system for designers to use in estimating manual assembly times [13,14]. To evaluate the effectiveness of this DFA method, the ease of assembly of a two-speed reciprocating power saw and an impact wrench was analyzed, and the products were then redesigned for easier assembly [15]. The initial design of the power saw (Fig. 7.13) had 41 parts and an

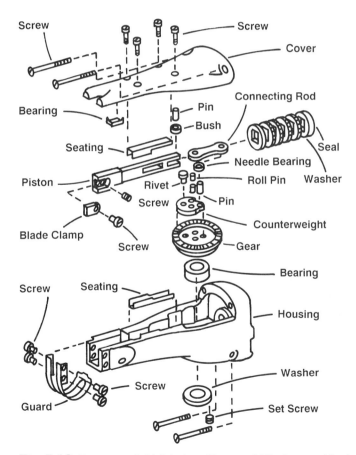

Fig. 7.13 Power saw (initial design: 41 parts, 6.37 min assembly time) (after Ref. 15).

estimated assembly time of 6.37 min. The redesign (Fig. 7.14) had 29 parts for a 29% reduction in part count and an estimated assembly time of 2.58 min for a 59% reduction in assembly time. The outcome of further analyses [15] was a more than 50% saving in assembly time, a significant reduction in parts count, and an anticipated improvement in product performance.

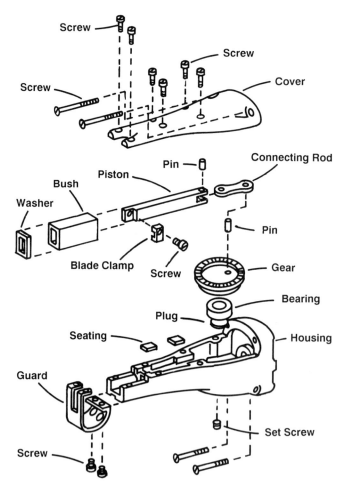

Fig. 7.14 Power saw (new design: 29 parts, 2.58 min assembly time) (after Ref. 15).

7.5 ASSEMBLY EFFICIENCY

An essential ingredient of the DFA method is the use of a measure of the assembly efficiency of a proposed design.

In general, the two main factors that influence the assembly cost of a product or subassembly are: (1) the total number of parts in a product, and (2) the ease of handling, insertion, and fastening of the parts. The term *assembly efficiency* is used to denote a figure obtained by dividing the theoretical minimum assembly time by the actual assembly time. The equation for calculating the manual assembly efficiency E_{ma} is

$$E_{ma} = N_{min} t_a / t_{ma} \qquad (7.1)$$

where N_{min} is the theoretical minimum number of parts, t_a the basic assembly time for one part, and t_{ma} is the estimated time to complete the assembly of the actual product. The basic assembly time is the average time for a part that presents no handling, insertion, or fastening difficulties.

The figure for the theoretical minimum number of parts represents an ideal situation in which separate parts are combined into a single part unless, as each part is added to the assembly, one of the following criteria is met:

1. The part moves relative to all other parts already assembled during the normal operating mode of the final product. (Small motions that can be accommodated by elastic hinges do not qualify.)

2. The part must be of a different material than all other parts assembled (for insulation, isolation, vibration damping, etc.).

3. The part must be separate from all other assembled parts; otherwise, the assembly of parts meeting one of the above criteria would be prevented.

It should be pointed out that these criteria are to be applied without taking into account general design requirements. For example, separate fasteners will not generally meet any of the above criteria and should always be considered for elimination. To be more specific, the designer considering the design of an automobile engine may feel that the bolts holding the cylinder head onto the engine block are necessary separate parts. However, they could be eliminated by combining the cylinder head with the block, an approach that is now being investigated by several manufacturers.

If applied properly, these criteria require the designer to consider means whereby the product can be simplified, and it is through this process that enormous improvements in manufacturability are often achieved. However, it is also necessary to be able to quantify the effects of changes in design schemes in terms of assembly time and costs. For this purpose, the DFA method incorporates a system for estimating assembly cost that, together with estimates of parts cost, will give the designer the information needed to make appropriate trade-off decisions.

7.6 CLASSIFiCATION SYSTEM FOR MANUAL HANDLING

The classification system for manual handling processes is a systematic arrangement of part features in order of increasing handling difficulty levels. The part features that affect manual handling time significantly are:

Size
Thickness
Weight
Nesting
Tangling
Fragility
Flexibility
Slipperiness
Stickiness
Necessity for using two hands
Necessity for using grasping tools
Necessity for optical magnification
Necessity for mechanical assistance

The classification system for manual handling processes, along with its associated definitions and corresponding time standards, is presented in Fig. 7.15. It can be seen that the classification numbers consist of two digits; each digit is assigned one of ten numerical symbols (0 to 9). The first digit of the coding system is divided into the following four main groups:

I.   First digit of 0–3   Parts of nominal size and weight that are easy to grasp and manipulate with one hand (without the aid of tools)

II. First digit of 4–7   Parts that require grasping tools to handle because of their small size

III. First digit of 8   Parts that severely nest or tangle in bulk

IV. First digit of 9   Parts that require two hands, two persons, or mechanical assistance in handling

Groups I and II are further subdivided into categories representing the amount of orientation required, based on the symmetry of the part.

The second digit of the handling code is based on flexibility, slipperiness, stickiness, fragility, and nesting characteristics of a part. The second digit also depends on the group divisions of the first digit in the following manner:

MANUAL HANDLING—ESTIMATED TIMES (seconds)

ONE HAND — parts can be grasped and manipulated by one hand without the aid of grasping tools

		parts are easy to grasp and manipulate					parts present handling difficulties (1)				
		thickness > 2 mm			thickness ≤ 2 mm		thickness > 2 mm			thickness ≤ 2 mm	
		size >15 mm	6 mm ≤ size ≤15 mm	size <6 mm	size >6 mm	size ≤6 mm	size >15 mm	6 mm ≤ size ≤15 mm	size <6 mm	size >6 mm	size ≤6 mm
		0	1	2	3	4	5	6	7	8	9
(α+β) < 360°	0	1.13	1.43	1.88	1.69	2.18	1.84	2.17	2.65	2.45	2.98
360° ≤ (α+β) < 540°	1	1.5	1.8	2.25	2.06	2.55	2.25	2.57	3.06	3	3.38
540° ≤ (α+β) < 720°	2	1.8	2.1	2.55	2.36	2.85	2.57	2.9	3.38	3.18	3.7
(α+β) = 720°	3	1.95	2.25	2.7	2.51	3	2.73	3.06	3.55	3.34	4

ONE HAND with GRASPING AIDS — parts can be grasped and manipulated by one hand but only with the use of grasping tools

		parts need tweezers for grasping and manipulation								parts need standard tools other than tweezers	parts need special tools for grasping and manipulation
		parts can be manipulated without optical magnification				parts require optical magnification for manipulation					
		parts are easy to grasp and manipulate		parts present handling difficulties (1)		parts are easy to grasp and manipulate		parts present handling difficulties (1)			
		thickness >0.25mm	thickness ≤0.25mm	thickness >0.25mm	thickness ≤0.25mm	thickness >0.25mm	thickness ≤0.25mm	thickness >0.25mm	thickness ≤0.25mm		
		0	1	2	3	4	5	6	7	8	9
0 ≤ β ≤ 180°, α ≤ 180°	4	3.6	6.85	4.35	7.6	5.6	8.35	6.35	8.6	7	7
β = 360°	5	4	7.25	4.75	8	6	8.75	6.75	9	8	8
0 ≤ β ≤ 180°, α = 360°	6	4.8	8.05	5.55	8.8	6.8	9.55	7.55	9.8	8	9
β = 360°	7	5.1	8.35	5.85	9.1	7.1	9.55	7.85	10.1	9	10

TWO HANDS for MANIPULATION — parts severely nest or tangle or are flexible but can be grasped and lifted by one hand (with the use of grasping tools if necessary) (2)

		parts present no additional handling difficulties					parts present additional handling difficulties (e.g. sticky, delicate, slippery, etc.) (1)				
		α ≤ 180°			α = 360°		α ≤ 180°			α = 360°	
		size >15 mm	6 mm ≤ size ≤15 mm	size <6 mm	size >6 mm	size ≤6 mm	size >15 mm	6 mm ≤ size ≤15 mm	size <6 mm	size >6 mm	size ≤6 mm
		0	1	2	3	4	5	6	7	8	9
	8	4.1	4.5	5.1	5.6	6.75	5	5.25	5.85	6.35	7

TWO HANDS or assistance required for LARGE SIZE — two hands, two persons or mechanical assistance required for grasping and transporting parts

		parts can be handled by one person without mechanical assistance								parts severely nest or tangle or are flexible (2)	two persons or mechanical assistance required for parts manipulation
		parts do not severely nest or tangle and are not flexible									
		part weight < 10 lb				parts are heavy (> 10 lb)					
		parts are easy to grasp and manipulate		parts present other handling difficulties (1)		parts are easy to grasp and manipulate		parts present other handling difficulties (1)			
		α≤180°	α=360°	α≤180°	α=360°	α≤180°	α=360°	α≤180°	α=360°		
		0	1	2	3	4	5	6	7	8	9
	9	2	3	2	3	3	4	4	5	7	9

Fig. 7.15 Classification, coding and data base for part features affecting manual handling time (© Boothroyd Dewhurst Inc., reproduced with permission).

I.	For a first digit of 0–3	The second digit classifies the size and thickness of the part.
II.	For a first digit of 4–7	The second digit classifies the part thickness, type of tool required for handling the part, and the necessity for optical magnification during the handling process.
III.	For a first digit of 8	The second digit classifies the size and symmetry of a part.
IV.	For a first digit of 9	The second digit classifies the symmetry, weight, and interlocking characteristics of parts in bulk.

7.7 CLASSIFICATION SYSTEM FOR MANUAL INSERTION AND FASTENING

The classification system for manual insertion and fastening processes is concerned with the interaction between mating parts as they contact and go together. Manual insertion and fastening consists of a finite variety of basic assembly tasks (peg-in-hole, screw, weld, rivet, force-fit, etc.) that are common to most manufactured products. The design features that significantly affect manual insertion and fastening times are:

Accessibility of assembly location
Ease of operation of assembly tool
Visibility of assembly location
Ease of alignment and positioning during assembly
Depth of insertion

The corresponding classification system and its associated definitions and time standards are presented in Fig. 7.16.

There were 100 code numbers in the original manual insertion and fastening coding system, as in the manual handling coding system. However, it was subsequently found [2] that certain categories of codes were not necessary in practice and are omitted in the latest version of the method. The two-digit code numbers range from 00 to 99. The first digit is divided into three main groups:

I.	First digit of 0–2	Part is not secured immediately after insertion.
II.	First digit of 3–5	Part secures itself or another part immediately after insertion.
III.	First digit of 9	Process involves parts that are already in place

Groups I and II are further subdivided into classes that consider the effect of obstructed access and/or restricted vision on assembly time.

MANUAL INSERTION — ESTIMATED TIMES (seconds)

PART ADDED but NOT SECURED

	after assembly no holding down required to maintain orientation and location (3)				holding down required during subsequent processes to maintain orientation or location (3)			
	easy to align and position during assembly (4)		not easy to align or position during assembly		easy to align and position during assembly (4)		not easy to align or position during assembly	
	no resistance to insertion	resistance to insertion (5)	no resistance to insertion	resistance to insertion (5)	no resistance to insertion	resistance to insertion (5)	no resistance to insertion	resistance to insertion (5)
	0	1	2	3	6	7	8	9
part and associated tool (including hands) can easily reach the desired location — 0	1.5	2.5	2.5	3.5	5.5	6.5	6.5	7.5
due to obstructed access or restricted vision (2) — 1	4	5	5	6	8	9	9	10
due to obstructed access and restricted vision (2) — 2	5.5	6.5	6.5	7.5	9.5	10.5	10.5	11.5

PART SECURED IMMEDIATELY

	no screwing operation or plastic deformation immediately after insertion (snap/press fits, circlips, spire nuts, etc.)		plastic deformation immediately after insertion						screw tightening immediately after insertion	
			plastic bending or torsion			rivetting or similar operation				
				not easy to align or position during assembly			not easy to align or position during assembly			
	easy to align and position with no resistance to insertion (4)	not easy to align or position during assembly and/or resistance to insertion (5)	easy to align and position during assembly (4)	no resistance to insertion	resistance to insertion (5)	easy to align and position during assembly (4)	no resistance to insertion	resistance to insertion (5)	easy to align and position with no torsional resistance (4)	not easy to align or position and/or torsional resistance (5)
	0	1	2	3	4	5	6	7	8	9
part and associated tool (including hands) can easily reach the desired location and the tool can be operated easily — 3	2	5	4	5	6	7	8	9	6	8
due to obstructed access or restricted vision (2) — 4	4.5	7.5	6.5	7.5	8.5	9.5	10.5	11.5	8.5	10.5
due to obstructed access and restricted vision (2) — 5	6	9	8	9	10	11	12	13	10	12

SEPARATE OPERATION

	mechanical fastening processes (part(s) already in place but not secured immediately after insertion)				non-mechanical fastening processes (part(s) already in place but not secured immediately after insertion)				non-fastening processes	
	none or localized plastic deformation				metallurgical processes					
						additional material required				
	bending or similar processes	rivetting or similar processes	screw tightening or other processes	bulk plastic deformation (large proportion of part is plastically deformed during fastening)	no additional material required, friction welding, etc.	soldering processes	weld/braze processes	chemical processes (e.g. adhesive bonding, etc.)	manipulation of parts or sub-assembly (e.g. orienting, fitting or adjustment of part(s), etc.)	other processes (e.g. liquid insertion, etc.)
	0	1	2	3	4	5	6	7	8	9
assembly processes where all solid parts are in place — 9	4	7	5	12	7	8	12	12	9	12

Fig. 7.16 Classification, coding, and data base for part features affecting insertion and fastening (© Boothroyd Dewhurst Inc., reproduced with permission).

The second digit of the assembly code is based on the following group divisions of the first digit:

I. For a first digit of 0–2   The second digit classifies the ease of engagement of parts and whether holding down is required to maintain orientation or location.

II. For a first digit of 3–5   The second digit classifies the ease of engagement of parts and whether the fastening operation involves a simple snap fit, screwing operation, or plastic deformation process.

III. For a first digit of 9   The second digit classifies mechanical, metallurgical, and chemical processes.

It can be seen in Figs. 7.15 and 7.16 that, for each two-digit code number, an average handling or insertion and fastening time is given. Thus, we have a set of time standards that can be used to estimate manual assembly times. These time standards were obtained from numerous experiments, some of which will now be described.

7.8 EFFECT OF PART SYMMETRY ON HANDLING TIME

One of the principal geometrical design features that affect the times required to grasp and orient a part is its symmetry. Assembly operations always involve at least two component parts: the part to be inserted and the part or assembly (receptacle) into which the part is inserted [16]. Orientation involves the proper alignment of the part to be inserted relative to the corresponding receptacle and can always be divided into two distinct operations: (1) alignment (rotation) of the axis of the part that corresponds to the axis of insertion, and (2) rotation of the part about this axis.

It is therefore convenient to define two kinds of symmetry for a part:

1. Alpha symmetry, which depends on the angle through which a part must be rotated about an axis perpendicular to the axis of insertion

2. Beta symmetry, which depends on the angle through which a part must be rotated about the axis of insertion

For example, a plain square prism that is to be inserted into a square hole would first have to be rotated about an axis perpendicular to the insertion axis, and since, with such a rotation, the maximum angle for orienting the prism is 180 deg, this can be termed 180-deg alpha symmetry. The square prism would then have to be rotated about the axis of insertion, and since the orientation of the prism about this axis would repeat every 90 deg, this implies a 90-deg beta symmetry. However, if the square prism were to be inserted in a circular hole, it would have 180-deg alpha symmetry and zero-deg beta symmetry. Figure 7.17 gives examples of the symmetry of simple-shaped parts.

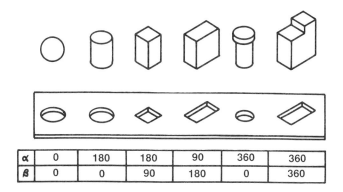

| α | 0 | 180 | 180 | 90 | 360 | 360 |
| β | 0 | 0 | 90 | 180 | 0 | 360 |

Fig. 7.17 Alpha and beta rotational symmetries for various parts.

A variety of predetermined time standard systems are currently used to establish assembly times in industry. In the development of these systems, several different approaches have been employed to determine relationships between the amount of rotation required to orient a part and the time required to perform that rotation. The two most commonly used systems are the methods time measurement (MTM) and work factor (WF) systems.

In the MTM system, the "maximum possible orientation" is employed, which is one-half the beta rotational symmetry of a part defined above [17]. The effect of alpha symmetry is not considered in this system. For practical purposes, the MTM system classifies the maximum possible orientation into three groups, namely; (1) symmetric, (2) semisymmetric, and (3) nonsymmetric [4]. Again, these terms refer only to the beta symmetry of a part.

In the WF system, the symmetry of a part is classified by the ratio of the number of ways the part can be inserted to the number of ways the part can be grasped preparatory to insertion [18]. In the example of a square prism to be inserted into a square hole, one particular end first can be inserted in four ways out of the eight ways it can be suitably grasped. Hence, on the average, one-half of the parts grasped require orientation, and this is defined in the WF system as a situation requiring 50% orientation [18]. Thus, in this system, account is taken of alpha symmetry, and some account is taken of beta symmetry. Unfortunately, these effects are combined in such a way that the classification can be applied to only a limited range of part shapes.

Numerous attempts were made to find a single parameter that would give a satisfactory relation between the symmetry of a part and the time required for orientation. It was found that the simplest and most useful parameter was the sum of the alpha and beta symmetries [6]. This parameter, which will be termed the *total angle of symmetry*, is therefore given by

$$\text{Total angle of symmetry} = \alpha + \beta \tag{7.2}$$

The effect of the total angle of symmetry on the time required to handle (grasp, move, orient, and place) a part is shown in Fig. 7.18. In addition, the shaded areas indicate the values of the total angle of symmetry that cannot exist. It is evident from these results that the symmetry of a part can be conveniently classified into five groups. However, the first group, which represents a sphere, is not generally of practical interest and, therefore, four groups are suggested that are employed in the coding system for part handling (Fig. 7.15).

Comparison of these experimental results with the MTM and WF orientation parameters showed that these parameters do not account properly for the symmetry of a part [6].

7.9 EFFECT OF PART THICKNESS AND SIZE ON HANDLING TIME

Two other major factors that affect the time required for handling during manual assembly are the thickness and the size of the part.

The thickness and size of a part are defined in a convenient way in the WF system, and these definitions have been adopted for the design for assembly

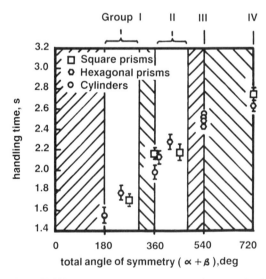

Fig. 7.18 Effect of symmetry on the time required to handle a part: (times are average for two individuals; shaded areas represent nonexisistent values of the total angle of symmetry.)

method. The thickness of a "cylindrical" part is its diameter whereas, for non-cylindrical parts, the thickness is defined as the maximum height of the part with its smallest dimension extending from a flat surface (Fig. 7.19). Cylindrical parts are defined as parts having cylindrical or other regular cross sections with five or more sides. When the diameter of such a part is greater than or equal to its length, the part is treated as noncylindrical. The reason for this distinction between cylindrical and noncylindrical parts in defining thickness is illustrated by the experimental curves shown in Fig. 7.19. It can be seen that parts with a "thickness" greater than 2 mm present no grasping or handling problems. However, for long cyclindrical parts, this critical value would have occurred at a value of 4 mm if the diameter had been used for the "thickness." Intuitively, we know that grasping a long cylinder 4 mm diameter is equivalent to grasping a rectangular part 2 mm thick if each is placed on a flat surface.

The size (also called the *major dimension*) of a part is defined as the largest nondiagonal dimension of the part's outline when projected on a flat surface. It is normally the length of the part. The effects of part size on handling time are shown in Fig. 7.20. Parts can be divided into four size categories as illustrated. Large parts involve little or no variation in handling time with changes in their size; the handling time for medium and small parts displays progressively greater sensitivity with respect to part size. Since the time penalty involved in handling very small parts is large and very sensitive to decreasing part size, tweezers are usually required to manipulate such parts.

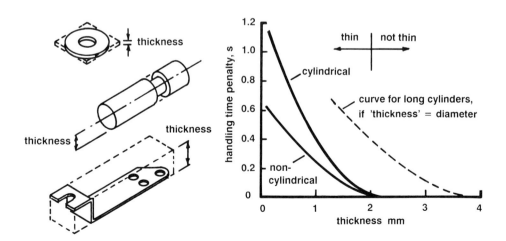

Fig. 7.19 Effect of part thickness on handling time.

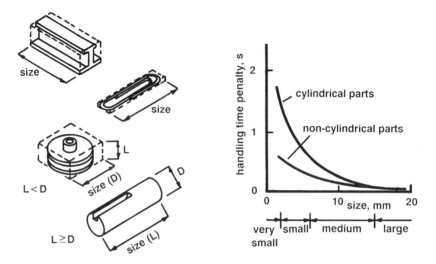

Fig. 7.20 Effect of part size on handling time.

7.10 EFFECT OF WEIGHT ON HANDLING TIME

Work has been carried out [19] on the effects of weight on the grasping, controlling, and moving of parts. The effect on grasping and controlling is found to be a time addition to the basic time for grasping and controlling, and the effect on moving is found to be a proportional increase. For the effect of weight on a part handled using one hand, the total adjustment t_{pw} to handling time can be represented by the following equation [4]:

$$t_{pw} = 0.0125W + 0.011Wt_h \tag{7.3}$$

where W(lb) is the weight of the part and t_h (sec) is the basic time for handling a "light" part when no orientation is needed and when it is to be moved a short distance.

An average value for t_h is 1.13 and, therefore, the total time penalty due to weight would be approximately $0.025W$.

If we assume that the maximum weight of a part to be handled using one hand is approximately 10–20 lb, the maximum penalty for weight is 0.25–0.5 sec and is a fairly small correction. Of course, this analysis does not take into account that larger parts will usually be moved greater distances, resulting in more significant time penalties.

7.11 PARTS REQUIRING TWO HANDS FOR MANIPULATION

Parts may require two hands for manipulation when:

1. The part is heavy
2. Very precise or careful handling is required
3. The part is large or flexible
4. The part does not possess holding features, thus making one-hand grasp difficult

Under these circumstances, a penalty is incurred because the second hand could be engaged in another operation, perhaps grasping another part. Experience shows that a penalty factor of 1.5 should be applied in these cases.

7.12 EFFECTS OF COMBINATIONS OF FACTORS

In the previous sections, various factors that affect manual handling times have been considered. However, it is important to realize that the penalties associated with each individual factor are not necessarily additive. For example, if a part requires additional time to move it from A to B, it can probably be oriented during the move. Therefore, it may be wrong to add the extra time for part size and an extra time for orientation to the basic handling time. The following gives some examples of results obtained when multiple factors are present.

7.13 EFFECT OF SYMMETRY FOR PARTS THAT SEVERELY NEST OR TANGLE AND MAY REQUIRE TWEEZERS FOR GRASPING AND MANIPULATION

A part may require tweezers when (Fig. 7.21):

1. Its thickness is so small that finger grasp is difficult
2. Vision is obscured and prepositioning is difficult because of its small size
3. It is undesirable to touch the part because of high-temperature, for example
4. Fingers cannot access the desired location.

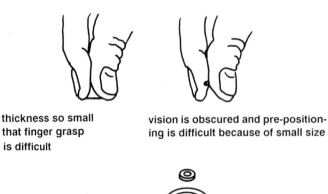

thickness so small   vision is obscured and pre-position-
that finger grasp   ing is difficult because of small size
is difficult

HOT   fingers cannot
                              access desired
                              location

undesirable to touch the part

Fig. 7.21 Examples of parts that may require tweezers for handling.

A part is considered to severely nest or tangle when an additional handling time of 1.5 sec or greater is required because of these factors. In general, two hands will be required to separate severely nested or tangled parts. Helical springs with open ends and widely spaced coils are examples of parts that severely nest or tangle.

Figure 7.22 shows how the time required for orientation is affected by the alpha and beta angles of symmetry for parts that nest or tangle severely and may require tweezers for handling.

In general, orientation using hands results in a smaller time penalty than orientation using tweezers; therefore, factors necessitating the use of tweezers should be avoided if possible.

7.14 EFFECT OF CHAMFER DESIGN ON INSERTION OPERATIONS

Two common assembly operations are the insertion of a peg (or shaft) into a hole and the placement of a part with a hole onto a peg.

The geometries of traditional conical chamfer designs are shown in Fig. 7.23. In Fig. 7.23a, which shows the design of a chamfered peg, d is the diameter of the peg, w_1 the width of the chamfer, and θ_1 the semiconical angle of

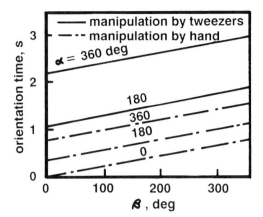

Fig. 7.22 Effect of symmetry on handling time when parts nest or tangle severely (disentangling time is not included).

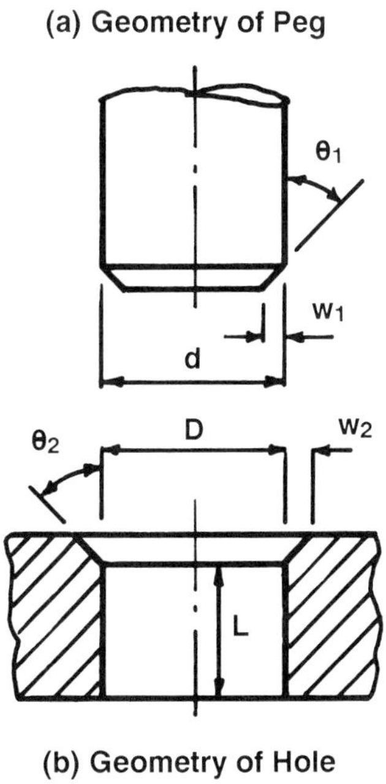

(a) Geometry of Peg

(b) Geometry of Hole

Fig. 7.23 Geometries of peg-and-hole.

the chamfer. In Fig. 7.23b, which shows the design of a chamfered hole, D is the diameter of the hole, w_2 the width of the chamfer, and θ_2 the semiconical angle of the chamfer. The dimensionless diametral clearance c between the peg and the hole is defined by

$$c = \frac{D - d}{D} \tag{7.4}$$

A typical set of results [10] showing the effects of various chamfer designs on the time taken to insert a peg in a hole are presented in Fig. 7.24. From these and other results, the following conclusions have been drawn: (1) for a given clearance, the difference in the insertion time for two different chamfer designs is always a constant; (2) a chamfer on the peg is more effective in reducing insertion time than the same chamfer on the hole; (3) the maximum width of the chamfer that is effective in reducing the insertion time for both the peg and the hole is approximately $0.1D$; (4) for conical chamfers, the most effective design provides chamfers on both the peg and the hole, and $w_1 = w_2 = 0.1D$ and $\theta_1 = \theta_2 < 45$ deg; (5) the manual insertion time is not sensitive to variations in the angle of the chamfer for the range $10\,\text{deg} < \theta < 50\,\text{deg}$; (6) a radiused or curved chamfer can have advantages over a conical chamfer for small clearances.

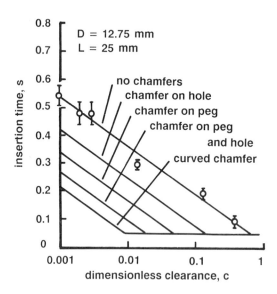

Fig. 7.24 Effect of clearance on insertion time (for clarity, experimental results are shown for only one case) (after Ref. 10).

It was learned from the peg insertion experiments [10] that the long manual insertion time for the peg and hole with a small clearance is probably due to the type of engagement occurring between the peg and the hole during the initial stages of insertion. Figure 7.25 shows two possible situations that will cause difficulties. In Fig. 7.25a, the two points of contact arising on the same circular cross section of the peg give rise to forces resisting the insertion. In Fig. 7.25b, the peg has become jammed at the entrance of the hole. An analysis was carried out to find a geometry that would avoid these unwanted situations. It showed that a chamfer conforming to a body of constant width (Fig. 7.26) is one of the designs having the desired properties. It was found that, for such a chamfer, the insertion time is independent of the dimensionless clearance c in the range $c > 0.01$. Therefore, the curved chamfer is the optimum design for the insertion operation (Fig. 7.24). However, since the manufacturing costs for curved chamfers would normally be greater than for conical chamfers, the modified chamfer would be worthy of consideration only for very small values of clearance when the significant reductions in insertion time might compensate for the higher cost. An interesting example of a curved chamfer is the geometry of a bullet. Such a design not only has aerodynamic advantages but also constitutes an ideal design for ease of insertion.

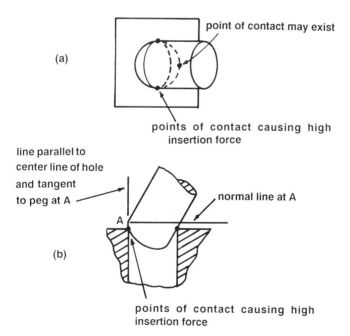

Fig. 7.25 Points of contact on chamfer and hole.

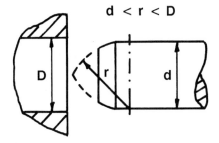

Fig. 7.26 Chamfer of constant width.

7.15 ESTIMATION OF INSERTION TIME

Empirical equations have been derived [10] to estimate the manual insertion time t_i for both conical chamfers and curved chamfers. For conical chamfers, the manual insertion time, t_i msec, is given by

$$t_i = \max(-80.5 \, \ell n \, c + f_1(w_1, w_2, D) + f_2(d_g) \\ + 1.4L + 289, 1.4L + 15) \tag{7.5}$$

where the function f_1 is given by

$$f_1 = \max\left[\frac{-1500w_1}{D} - \frac{1200w_2}{D}, -270\right] \tag{7.6}$$

the function f_2 is given by

$$f_2 = 0.602d_g^2 - 33.2d_g \tag{7.7}$$

and where d_g is the grip size (mm) and L is the insertion depth (mm). For modified curved chamfers, the insertion time is given by

$$t_i = 1.4L + 15 \tag{7.8}$$

7.16 AVOIDING JAMS DURING ASSEMBLY

Parts with holes that must be assembled onto a peg can easily jam if they are not dimensioned carefully. This problem is typical of assembling a washer on a bolt.

In analyzing a part assembled on a peg, [8] the hole diameter can be taken to be one unit; all other length dimensions are then measured relative to this unit and are dimensionless (Fig. 7.27). The peg diameter is $1 - c$, where c is

(a) **Part Assembled on Peg**   (b) **Part Wedged on Peg**

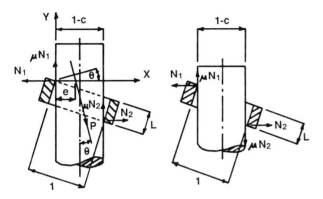

Fig. 7.27 Geometry of part and peg.

the dimensionless diametral clearance between the two mating parts. The resultant force in the assembly operation is denoted by P. The line of action of P intercepts the X axis at e, 0. If the following equation is satisfied, the part will slide freely down the peg:

$$P \cos \theta > \mu(N_1 + N_2) \tag{7.9}$$

By resolving forces horizontally,

$$P \sin \theta + N_2 - N_1 = 0 \tag{7.10}$$

and, by taking moments about $(0,0)$,

$$\{[1 + L^2 - (1 - c)^2]^{1/2} + \mu(1 - c)\}N_2 - eP \cos \theta = 0 \tag{7.11}$$

From Eqs. (7.9), (7.10), and (7.11),

$$\left(\frac{2\mu e}{q} - 1 \right) \cos \theta + \mu \sin \theta < 0 \tag{7.12}$$

where

$$q = [1 + L^2 - (1 - c)^2]^{1/2} + \mu(1 - c)$$

Thus, when $e = 0$ and $\cos \theta > 0$, the condition

$$\tan \theta < \frac{1}{\mu} \tag{7.13}$$

ensures free sliding. If $e = 0$ and $\cos \theta$ is less than 0, then the condition becomes

$$\tan \theta > \frac{1}{\mu} \qquad (7.14)$$

In the case in which $\theta = 0$ (the assembly force is applied vertically), Eq. (7.12) yields

$$2\mu e < q \qquad (7.15)$$

or

$$e = \frac{m}{2}(1 - c) \qquad (7.16)$$

where m is a positive number. Substituting Eq. (7.16) into Eq. (7.15) gives

$$1 + L^2 > (1 - c)^2[\mu^2(m - 1)^2 + 1] \qquad (7.17)$$

When $m = 1$, the force is applied along the axis of the peg. Because $(1 + L^2)$ must always be larger than $(1 - c)^2$, the parts will never jam under these circumstances.

It also is necessary to consider whether the part can be rotated and wedged on the peg. If the net moment of the reaction forces at the contact points is in the direction that rotates the part from the wedged position, then the part will free itself. Thus, for the part to free itself,

$$1 + L^2 > (1 - c)^2(\mu^2 + 1) \qquad (7.18)$$

Comparing Eq. (7.17) with Eq. (7.18) shows that the condition for the part to wedge without freeing itself is produced when $m = 2$ in Eq. (7.17).

7.17 REDUCING DISK-ASSEMBLY PROBLEMS

When an assembly operation calls for the insertion of a disk-shaped part into a hole, jamming or hang-up is a common problem. Special handling equipment can prevent jams, but a simpler, less costly solution is to analyze carefully all part dimensions before production begins.

Again, the diameter of the hole is one unit; all other dimensions are measured relative to this unit and are dimensionless (Fig. 7.28). The disk diameter is $1 - c$, where c is the dimensionless diametral clearance between the mating parts, P the resultant force in the assembly operation, and μ the coefficient of friction.

When a disk with no chamfer is inserted into a hole, the condition for free sliding can be determined by

$$L^2 > \mu^2 + 2c - c^2 \qquad (7.19)$$

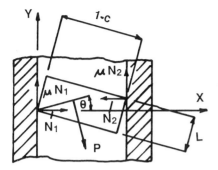

Fig. 7.28 Geometry of disk and hole.

If c is very small, then Eq. (7.19) can be expressed as

$$L > \mu + \frac{c}{\mu} \tag{7.20}$$

If the disk is very thin, that is, if

$$(1 - c)^2 + L^2 < 1 \tag{7.21}$$

the disk can be inserted into the hole by keeping its circular cross section parallel to the wall of the hole and reorienting it when it reaches the bottom of the hole.

7.18 EFFECTS OF OBSTRUCTED ACCESS AND RESTRICTED VISION ON INSERTION OF THREADED FASTENERS OF VARIOUS DESIGNS

Considerable experimental work has been conducted on the time taken to insert threaded fasteners of different types and under a variety of conditions. First, let us consider the time taken to insert a machine screw and engage the threads; Fig. 7.29a shows the effects of the shape of the screw point and hole entrance when the assembly worker cannot see the operation and when various levels of obstruction are present. When the distance from the obstructing surface to the hole center was greater than 16 mm, the surface had no effect on the manipulations, and the restriction of vision was the only factor. Under these circumstances, the standard screw inserted into a recessed hole gave the shortest time. For a standard screw with a standard hole, an additional 2.5 sec was required. When the hole was closer to the wall, thereby inhibiting the manipulations, a further time of 2–3 sec was necessary.

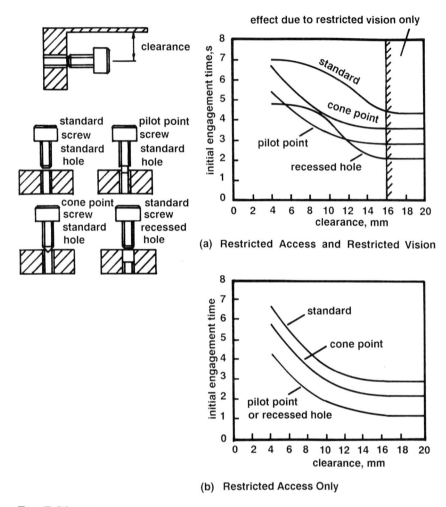

Fig. 7.29 Effects of restricted access and restricted vision on initial engagement of screws.

Figure 7.29b shows the results obtained under similar conditions but with the worker's vision unrestricted. Comparison with the previous results indicates that restriction of vision had little effect when access was obstructed. This was because the proximity of the obstructing surface allowed tactile sensing to take the place of sight. When the obstruction was removed, however, restricted vision could account for up to 1.5 sec additional time.

Once the screw threads are engaged, the assembly worker must grasp the necessary tool, engage it with the screw, and perform sufficient rotations to tighten the screw. Figure 7.30 shows the total time for these operations for a variety of screw head designs and for both hand-operated and power tools. There was no restriction on tool operation for any of these situations. Finally, Fig. 7.31 shows the time to turn down a nut using a variety of hand-operated tools, where the operation of the tools was obstructed to various degrees. It can be seen that the penalties for a box-end wrench are as high as 4 sec per revolution when obstructions are present. However, when planning the design of a new product, the designer does not normally consider the type of tool used and can reasonably expect that the best tool for the job will be selected. In the present case, this is either the nut driver or the socket ratchet wrench.

7.19 EFFECTS OF OBSTRUCTED ACCESS AND RESTRICTED VISION ON POP-RIVETING OPERATIONS

Figure 7.32 summarizes the results of experiments [11] on the time taken to perform pop-riveting operations. In the experiments, the average time taken to pick up the tool, change the rivet, move the tool to the correct location, insert the rivet, and return the tool to its original location was 7.27 sec. Figure 7.32a summarizes the combined effects of obstructed access and restricted vision, and Fig. 7.32b shows the effects of obstructed access alone. In the latter case,

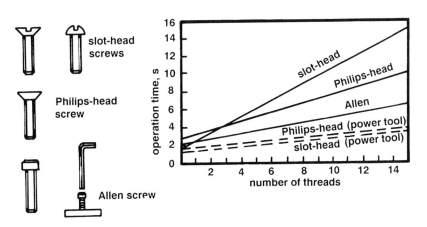

Fig. 7.30 Effect of number of threads on time to pick up the tool, engage the screw, tighten the screw, and replace the tool.

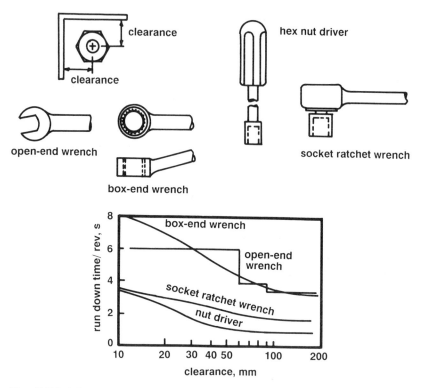

Fig. 7.31 Effect of obstructed access on time to tighten a nut.

time penalties of up to 1 sec can be incurred although, unless the clearances are quite small, the penalties are negligible. With restricted vision present, much higher penalties, on the order of 2–3 sec, were obtained.

7.20 EFFECTS OF HOLDING DOWN

Holding down is required when parts are unstable after insertion or during subsequent operations. It is defined as a process that, if necessary, maintains the position and orientation of parts already in place prior to, or during, subsequent operations. The time taken to insert a peg vertically through holes in two or more stacked parts can be expressed as the sum of a basic time t_b and a time penalty t_p. The basic time is the time to insert the peg when the parts are prealigned and self-locating, as shown in Fig. 7.33a, and can be expressed [12] as

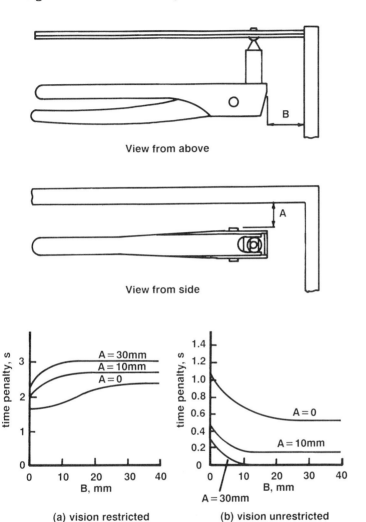

Fig. 7.32 Effects of obstructed access and restricted vision on the time to insert a pop rivet (after Ref. 11).

$$t_b = -0.07 \ln c - 0.1 + 3.7L + 0.75d_g \qquad (7.22)$$

where $c = (D - d)/D$ and is the dimensionless clearance $(0.1 \geqslant c \geqslant 0.001)$, L is the insertion depth in meters, and d_g is the grip size in meters $(0.1\,m \geqslant d_g \geqslant 0.01\,m)$. For example,

$$D = 20 \text{ mm} \qquad d = 19.6 \text{ mm} \qquad c = \frac{(D - d)}{D} = \frac{(20 - 19.6)}{20} = 0.02$$

$$L = 100 \text{ mm} = 0.10 \text{ m (assume 3 interfaces)} \qquad d_g = 40 \text{ mm} = 0.04 \text{ m}$$

then,

$$t_b = -0.07 \, \ell n \, c - 0.1 + 3.7L + 0.75 d_g$$

$$= -0.07 \times \ell n \, 0.02 - 0.1 + 3.7 \times 0.10 + 0.75 \times 0.04$$

$$= 0.27 - 0.1 + 0.37 + 0.03$$

$$= 0.57 \text{ sec}$$

The graphs presented in Figs. 7.33 and 7.34 will allow the time penalty t_p to be determined for three conditions:

1. When easy-to-align parts have been aligned and require holding down (Fig. 7.33b)
2. When difficult-to-align parts have been aligned and require holding down (Fig. 7.33c)
3. When difficult-to-align parts require alignment and holding down (Fig. 7.34)

For the example given above in which $t_b = 0.57$ sec, the time penalty $t_p = 0.09$ sec for the conditions of Fig. 7.33b, the time penalty $t_p = 0.155$ sec for the conditions of Fig. 7.33c, and $t_p = 3$ sec for the conditions of Fig. 7.34.

7.21 MANUAL ASSEMBLY DATA BASE AND DESIGN DATA SHEETS

The above sections have presented a selection of the results of some of the analyses and experiments conducted during the early phases of development of the design for assembly method.

For the development of the classification schemes and time standards presented earlier, it was necessary to obtain an estimate of the average time, in seconds, to complete the operation for all the parts falling within each classification or category. For example, the top item in the left-hand column in Fig. 7.15 (code 00) gives a figure of 1.13 for the average time to grasp, orient, and move a part that

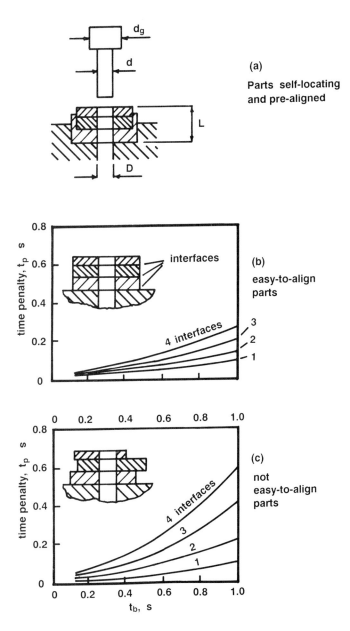

Fig. 7.33 Effects of holding down on insertion time (after Ref. 12).

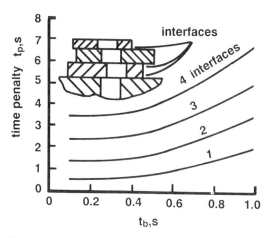

Fig. 7.34 Effects of holding down and realignment on insertion time for difficult-to align parts (after Ref. 12).

1. Can be grasped and manipulated with one hand
2. Has a total symmetry angle of less than 360 deg (a plain cylinder, for example)
3. Is larger than 15 mm
4. Has a thickness greater than 2 mm
5. Has no handling difficulties, such as flexibility, tendency to tangle or nest, etc.

Clearly, a wide range of parts will fall within this category, and their handling times will vary somewhat. The figure presented is only an average figure for the range of parts.

To illustrate the type of problem that can arise through the use of the group technology coding or classification scheme employed in the "Design for Assembly Handbook," we can consider the assembly of a part having a thickness of 1.9 mm. We shall assume that, except for its thickness of less than 2 mm, the part would be classified as code 00 (Fig. 7.15). However, because of the thickness of the part, the appropriate code would be 03, and the estimated handling time would be 1.69 sec instead of 1.13 sec, which represents a time penalty of 0.56 sec. Turning now to the results of experiments for the effect of thickness (Fig. 7.19), we can see that, for a cylindrical part, the actual time penalty is on the order of only 0.01–0.02 sec. We would therefore expect an error in our results of about 50%.

Experience has shown that, under normal circumstances, these errors tend to cancel—with some parts, the error results in an overestimate of time and, with some, an underestimate. However, if an assembly contains a large number of identical parts, care must be taken to check whether the part characteristics fall close to the limits of the classification; if they do, then the detailed results presented above should be consulted.

7.22 APPLICATION OF THE DFA METHODOLOGY

To illustrate how DFA is applied in practice, we shall consider the controller assembly shown in Fig. 7.35. The assembly of this product first involves securing a series of assemblies to the metal frame using screws, connecting these assemblies together in various ways, and then securing the resulting assembly into the plastic cover, again using screws. An undesirable feature of the design of the plastic cover is that the small subassemblies must be fastened to the metal frame before the metal frame can be secured to the plastic cover.

Figure 7.36 shows a completed worksheet analysis for the controller in the form of a tabulated list of operations and the corresponding assembly times and costs. Each assembly operation is divided into handling and insertion, and the corresponding times and two-digit code numbers for each process are given. Assembly starts by placing the pressure regulator (a purchased item) upside down into a fixture. The metal frame is placed onto the projecting spindle of the pressure regulator and secured with the nut. The resulting assembly is then turned over in the fixture to allow for the addition of other items to the metal frame.

Next, the sensor and the strap are placed and held in position while two screws are installed. Clearly, the holding of these two parts and the difficulty of the screw insertions will impose time penalties on the assembly process.

After applying tape to the thread on the sensor, the adapter nut can be screwed into place. Then, one end of the tube assembly is screwed to the threaded extension on the pressure regulator and the other end to the adapter nut. Clearly, both of these are difficult and time-consuming operations.

The PCB assembly is now positioned and held in place while two screws are installed; after which its connector is snapped into the sensor and the earth lead is snapped into place.

The whole assembly must be turned over once again to allow for the positioning and holding of the knob assembly while the screw fastening operation can be carried out. Finally, the plastic cover is placed in position and the entire assembly turned over for the third time to allow the three screws to be inserted. It should be noted that access for the insertion of these screws is very restricted.

It is clear from this description of the assembly sequence that many aspects of the design could be improved. However, a step-by-step analysis of each

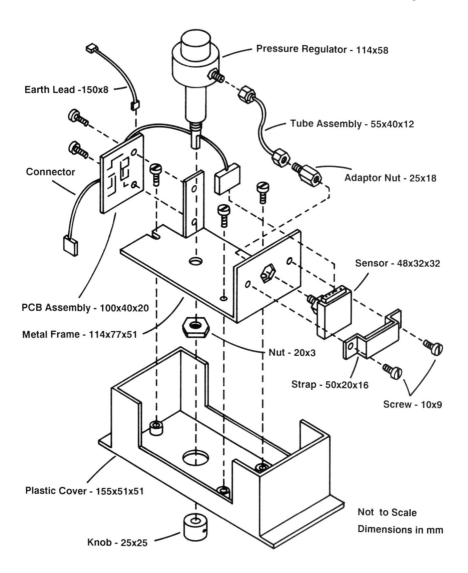

Pressure Regulator - 114x58

Earth Lead -150x8

Tube Assembly - 55x40x12

Connector

Adaptor Nut - 25x18

Sensor - 48x32x32

PCB Assembly - 100x40x20

Metal Frame - 114x77x51

Nut - 20x3

Strap - 50x20x16

Screw - 10x9

Plastic Cover - 155x51x51

Not to Scale

Dimensions in mm

Knob - 25x25

Fig. 7.35 Controller assembly.

MANUAL - BENCH ASSEMBLY	Manual handling code			Manual Insertion code		Total oper'n time RP*(TH+TI)	Figures for min. parts		Operator rate OP: 30.00 $/hr 0.83 c/s
Name of Assembly - $MAIN SUB	No. of items	Handling time per item (s)		Insertion time per item (s)		Total oper'n cost-cents TA*OP			
Item Name: Part, Sub or Pcb assembly									Description
No. or Operation	RP	HC	TH	IC	TI	TA	CA	NM	
1 $pressure regulator	1	30	1.95	00	1.5	3.5	2.9	1	place in fixture
2 metal frame	1	30	1.95	06	5.5	7.4	6.2	1	add
3 nut	1	00	1.13	39	8.0	9.1	7.6	0	add & screw fasten
4 Reorientation	1	-	-	98	9.0	9.0	7.5	-	reorient & adjust
5 $sensor	1	30	1.95	08	6.5	8.4	7.0	1	add
6 strap	1	20	1.80	08	6.5	8.3	6.9	0	add & hold down
7 Screw	2	11	1.80	39	8.0	19.6	16.3	0	add & screw fasten
8 Apply tape	1	-	-	99	12.0	12.0	10.0	-	special operation
9 adaptor nut	1	10	1.50	49	10.5	12.0	10.0	0	add & screw fasten
10 tube assembly	1	91	3.00	10	4.0	7.0	5.8	0	add & screw fasten
11 Screw fastening	1	-	-	92	5.0	5.0	4.2	-	standard operation
12 &PCB ASSEMBLY	1	83	5.60	08	6.5	12.1	10.1	1	add & hold down
13 Screw	2	11	1.80	39	8.0	19.6	16.3	0	add & screw fasten
14 connector	1	30	1.95	31	5.0	6.9	5.8	0	add & snap fit
15 earth lead	1	83	5.60	31	5.0	10.6	8.8	0	add & snap fit
16 Reorientation	1	-	-	98	9.0	9.0	7.5	-	reorient & adjust
17 $knob assembly	1	30	1.95	08	6.5	8.4	7.0	1	add & screw fasten
18 Screw fastening	1	-	-	92	5.0	5.0	4.2	-	standard operation
19 plastic cover	1	30	1.95	08	6.5	8.4	7.0	0	add & hold down
20 reorientation	1	-	-	98	9.0	9.0	7.5	-	reorient & adjust
21 screw	3	11	1.80	49	10.5	36.9	30.8	0	add & screw fasten

Fig. 7.36 Completed worksheet analysis for the controller assembly.

operation is necessary before changes to simplify the product structure and reduce assembly difficulties can be identified and quantified. First, we shall look at how the handling and insertion times are established. The addition of the strap to the metal frame will be considered by way of example. This operation is item 5 on the worksheet, and the line of information is completed as follows:

Number of Items, RP

There is one strap.

Handling Code, HC

The insertion axis for the strap is horizontal in Fig. 7.35, and the strap can be inserted only one way along this axis so that the alpha angle of symmetry is 360 deg. If the strap is rotated about the axis of insertion, it will repeat its orientation every 180 deg, which is, therefore, the beta angle of symmetry. Thus, the total angle of symmetry is 540 deg. Referring to the data base for handling time (Fig. 7.15), since the strap can be grasped and manipulated using one hand without the aid of tools and since alpha plus beta is 540 deg, the first digit of the handling code is 2. The strap presents no handling difficulties (can be grasped and separated from bulk easily), its thickness is greater than 2 mm, and its size is greater than 15 mm; therefore, the second digit is 0, giving a handling code of 20.

Handling Time Per Item, TH

A handling time of 1.8 sec corresponds to a handling code of 20 (Fig. 7.15).

Insertion Code, IC

The strap is not secured as part of the insertion process, and since there is no restriction to access or vision, the first digit of the insertion code is 0 (Fig. 7.16). Holding down is necessary while subsequent operations are carried out, and the strap is not easy to align because no features are provided to facilitate alignment of the screw holes. There will be no resistance to insertion and, therefore, the second digit will be 8, giving an insertion code of 08.

Insertion Time Per Item, TI

An insertion time of 6.5 sec corresponds to an insertion code of 08 (Fig. 7.16).

Total Operation Time, TA

This is the sum of the handling and insertion times multiplied by the number of items, i.e., RP(TH + TI). For the strap, the total operation time is therefore 8.3 sec.

Total Operation Cost, CA

The cost of manual assembly depends on the manual assembly worker's rate. This rate should include overheads and is usually referred to as the *burdened rate*. Rates vary from region to region and from factory to factory, and so it is usually necessary to determine the appropriate rate for a particular company. However, a typical figure would be $30/hr, which converts to 0.83 cents/sec. The total operation cost is now obtained by multiplying the total operation time by the assembly worker rate and, for the strap, this would be $8.3 \times 0.83 = 6.9$ cents.

Figures For Minimum Parts, NM

As explained earlier, the establishment of a theoretical minimum part count is the most powerful way to identify possible simplifications in the product structure. For the strap, the three criteria for separate parts are applied after the pressure regulator, the metal frame, the nut, and the sensor have been assembled:

1.   The strap does not move relative to these parts, and so it could theoretically be combined with any of them.

2.   The strap does not have to be of a different material—in fact, it could be of the same plastic material as the body of the sensor and therefore take the form of two lugs with holes projecting from the body. At this point in the analysis, the designer would probably determine that, since the sensor is a purchased stock item, its design could not be changed. However, it is important to ignore these economic considerations at this stage and consider only theoretical possibilities.

3.   The strap clearly does not have to be separate from the sensor in order to allow assembly of the sensor and, therefore, none of the three criteria are met, and the strap becomes a candidate for elimination. For the strap, a zero is placed in the column for minimum parts.

7.23 RESULTS OF THE ANALYSIS

Once the analysis is complete for all operations, the appropriate columns can be summed. Thus, for the controller, the total number of parts and subassemblies is 19, and there are 6 additional operations. The total assembly time is 227 sec, the corresponding assembly cost is $1.90, and the theoretical minimum number of items is 5.

A manual assembly design efficiency is now obtained using Eq. (7.1). In this equation, t_a is the basic assembly (handling and insertion) time for one part and can be taken as 3 sec on average. Thus, efficiency,

$$E_{\text{ma}} = \frac{N_{\text{min}} t_a}{t_{\text{ma}}}$$

$$= \frac{5 \times 3}{227.4}$$

$$= 0.07 \text{ or } 7\%$$

The high cost processes should now be identified, especially those associated with the installation of parts that do not meet any of the criteria for separate parts. From the worksheet results (Fig. 7.36), it can be seen that attention should clearly be paid to combining the plastic cover with the metal frame. This would eliminate the assembly operation for the cover, the three screws, and the reorientation operation, representing a total time saving of 54.3 sec, which constitutes 24% of the total assembly time. Of course, the designer must check that the cost of the combined plastic cover and frame is less than the total cost of the individual items.

A summary of the items that can be identified for elimination or combination and the appropriate assembly time savings are presented in Table 7.1.

We have now identified design changes that could result in saving at least 149.4 sec of assembly time, which constitutes 66% of the total. In addition,

Table 7.1 Possible Design Changes for the Controller

Design change	Items	Time saving, sec
1. Combine plastic cover with frame and eliminate 3 screws and reorientation	19,20,21	54.3
2. Eliminate strap and 2 screws (snaps in plastic frame to hold sensor, if necessary)	6,7	27.9
3. Eliminate screws holding PCB assembly (provide snaps in plastic frame)	13	19.6
4. Eliminate 2 reorientations	4, 16	18.0
5. Eliminate tube assembly and 2 screwing operations (screw adapter nut and sensor direct to the pressure regulator)	10, 11	12.0
6. Eliminate earth lead (not necessary with plastic frame)	15	10.6
7. Eliminate connector (Plug sensor into PCB)	14	7.0

several items of hardware would be eliminated, resulting in reduced part costs. Figure 7.37 shows a conceptual redesign of the controller in which all the proposed design changes have been made, and Fig. 7.38 presents the corresponding revised worksheet. The total assembly time is now 84 sec, and the assembly efficiency is increased to 18%, a fairly respectable figure for this type of assembly. Of course, the designer or design team must now consider the technical and economic consequences of the proposed designs.

First, there is the effect on the cost of the parts. However, experience shows, and this example would be no exception, that the saving from parts cost reduction would be greater than the saving in assembly costs which, in this case, is $1.20.

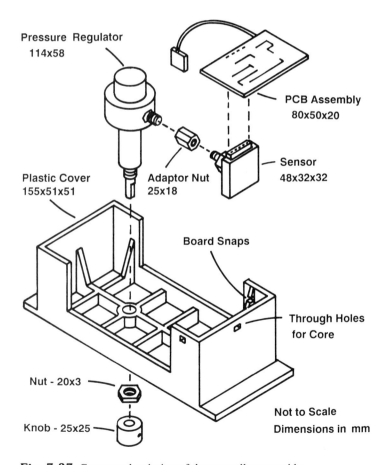

Fig. 7.37 Conceptual redesign of the controller assembly.

MANUAL - BENCH ASSEMBLY	No. of items	Manual handling code	Handling time per item (s)	Manual Insertion code	Insertion time per item (s)	Total oper'n time RP*(TH+TI)	Total oper'n cost-cents TA*OP	Figures for min. parts	Operator rate OP: 30.00 $/hr 0.83 c/s	SUB ASSEMBLY OR PART COSTS	
Name of Assembly - $MAIN SUB										Total item cost $	Total tooling cost k$
No. Item Name: Part, Sub or Pcb assembly or Operation	RP	HC	TH	IC	TI	TA	CA	NM	Description	CT	CC
1 $pressure regulator	1	30	1.95	00	1.5	3.5	2.9	1	place in fixture	10.46	0.0
2 plastic cover	1	30	1.95	06	5.5	7.4	6.2	1	add & hold down	0.00	0.0
3 nut	1	00	1.13	39	8.0	9.1	7.6	0	add & screw fasten	0.20	0.0
4 $knob assembly	1	30	1.95	08	6.5	8.4	7.0	1	add & screw fasten	-	-
5 Screw fastening	1	-	-	92	5.0	5.0	4.2	-	standard operation	-	-
6 Reorientation	1	-	-	98	9.0	9.0	7.5	-	reorient & adjust	-	-
7 Apply tape	1	-	-	99	12.0	12.0	10.0	-	special operation	-	-
8 adaptor nut	1	10	1.50	49	10.5	12.0	10.0	0	add & screw fasten	0.30	0.0
9 $SENSOR	1	30	1.95	39	8.0	9.9	8.3	1	add & screw fasten	1.50	0.0
10 &PCB ASSEMBLY	1	83	5.60	30	2.0	7.6	6.3	1	add & snap fit	0.00	0.0

Fig. 7.38 Completed anaysis for the controller assembly redesign.

It should be realized that the documented savings in materials, manufacturing, and assembly represent direct costs. To obtain a true picture, overheads must be added, and these can often amount to 200% or more. In addition, there are other savings more difficult to quantify. For example, when a part such as the metal frame is eliminated, all associated documentation, including part drawings, is also eliminated. Also, the part cannot be misassembled or fail in service, factors that lead to improved reliability, maintainability, and quality of the product. It is not surprising, therefore, that many U.S. companies have been able to report annual savings measured in millions of dollars as a result of the application of the DFA analysis method described here.

7.24 FURTHER GENERAL DESIGN GUIDELINES

Some guidelines or design rules for the manual handling and insertion of parts were listed earlier. However, it is possible to identify a few more general guidelines that arise particularly from the application of the minimum parts criteria, many of which found application in the analysis of the controller.

1. Avoid Connections:
If the only purpose of a part or assembly is to connect A to B, then try to locate A and B at the same point.

Figure 7.39 illustrates this guideline. Here, the two connected assemblies are rearranged to provide increasing assembly and manufacturing efficiency. Also, two practical examples occurred during the analysis of the controller when it was found that the entire tube assembly could be eliminated and that

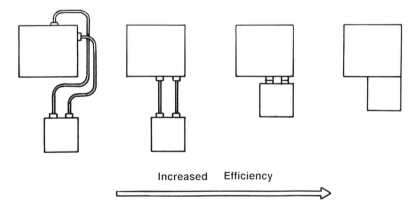

Increased   Efficiency

Fig. 7.39 Rearrangement of connected items to improve assembly efficiency and reduce costs.

the wires from the PCB assembly to the connector were not necessary (Fig. 7.35).

    2.   Design so that access for assembly operations is not restricted.

Figure 7.40 shows two alternative design concepts for a small assembly. In the first concept, the installation of the screws would be very difficult because of the restricted access within the box-shaped base part. In the second concept, access is relatively unrestricted because the assembly is built up on the flat base part.

    An example of this type of problem occurred in the controller analysis when the screws holding the metal frame assembly in the plastic cover were installed (item 21, Fig. 7.36).

    3.   Avoid adjustments

Figure 7.41 shows two parts of different materials secured by two screws in such a way that adjustment of the overall length of the assembly is necessary. If the assembly was replaced by one part manufactured from the more expen-

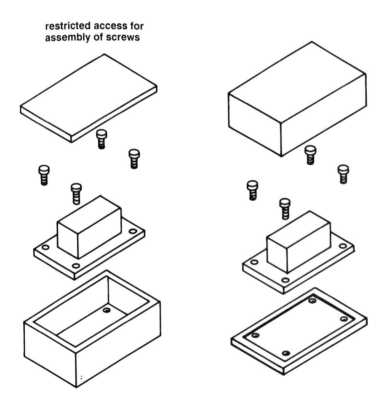

Fig. 7.40 Design concept to provide easier access during assembly.

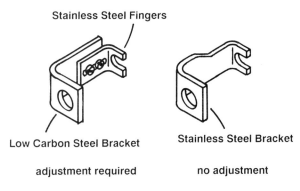

Fig. 7.41 Design to avoid adjustment during assembly.

sive material, difficult and costly operations would be avoided. These savings would probably more than offset the increase in material costs.

4. Use kinematic design principles

There are many ways in which the application of kinematic design principles can reduce manufacturing and assembly costs. Invariably, when located parts are overconstrained, it is necessary either to provide a means of adjustment for the constraining items or to employ more accurate machining operations. Figure 7.42a shows an example in which six point constraints are used to locate the square block in the plane of the page. Each constraint requires adjustment. According to kinematic design principles, only three point constraints are needed, together with closing forces. Clearly, the redesign shown in Fig. 7.42b is simpler, requiring fewer parts, fewer assembly operations, and less adjustment.

In many circumstances, designs in which overconstraint is involved result in redundant parts. In the design involving overconstraint in Fig. 7.43, one of

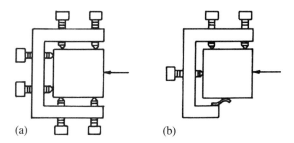

Fig. 7.42 Overconstraint leads to unnecessary complexity in product design: (a) overconstrained design; (b) sound kinematic design.

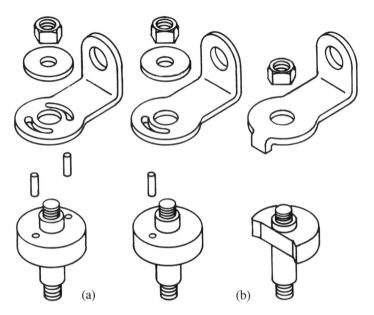

Fig. 7.43 Overconstraint leads to redundancy of parts: (a) overconstrained; (b) kinematically sound.

the pins is redundant. However, the application of the minimum parts criteria to the design with a single pin would suggest combining the pin with one of the major parts and combining the washer with the nut.

REFERENCES

1. Militzer, R.W., "Needed: Manufacturing Input to Product Design," President's Message, *Manufacturing Engineering*, Vol. 82, Feb. 1979.
2. Boothroyd, G., and Dewhurst, P., "Product Design for Assembly," Boothroyd Dewhurst Inc., Wakefield, R I., 1986.
3. Boothroyd, G., "Design for Economic Manufacture," *Annals of the CIRP*, Vol. 28, No. 1, 1979, p. 345.
4. Yoosufani, Z., and Boothroyd, G., "Design of Parts for Ease of Handling," Report #2, Dept. of Mechanical Engineering, Univ. of Massachusetts, Amherst, Sept, 1978.
5. Boothroyd, G., "Design for Manual Handling and Assembly," Report #4, Dept. of Mechanical Engineering, Univ. of Massachusetts, Amherst, Sept. 1979.
6. Yoosufani, Z., Ruddy, M., and Boothroyd, G., "Effect of Part Symmetry on Manual Assembly Times," *Journal of Manufacturing Systems*, Vol. 2, No. 2, 1983, pp. 189–195.
7. Seth, B., and Boothroyd, G., "Design for Manual Handling," Report #9, Dept. of Mechanical Engineering, Univ. of Massachusetts, Amherst, Jan. 1979.

8. Ho, C., and Boothroyd, G., "Avoiding Jams During Assembly," *Machine Design*, Technical Brief, Jan. 25, 1979.

9. Ho, C., and Boothroyd, G., "Reducing Disk-Assembly Problems," *Machine Design*, Technical Brief, March 8, 1979.

10. Ho, C., and Boothroyd, G., "Design of Chamfers for Ease of Assembly," *Proceedings of the 7th North American Metalworking Conference*, S.M.E. May 1979, p. 345.

11. Fujita, T., and Boothroyd, G., "Data Sheets and Case Study for Manual Assembly," Report #16, Dept. of Mechanical Engineering, Univ. of Massachusetts, Amherst, April 1982.

12. Yang, S.C., and Boothroyd, G., "Data Sheets and Case Study for Manual Asssembly," Report #15, Dept. of Mechanical Engineering, Univ. of Massachusetts, Amherst, Dec. 1981.

13. Dvorak, W.A., and Boothroyd, G., "Design for Assembly Handbook," Report #11, Dept. of Mechanical Engineering, Univ. of Massachusetts, Amherst, Dec. 1980.

14. De Lisser, W.A., and Boothroyd, G., "Analysis of Product Designs for Ease of Manual Assembly—A Systematic Approach," Report #17, Dept. of Mechanical Engineering, Univ. of Massachusetts, Amherst, May 1982.

15. Ellison, B., and Boothroyd, G., "Applying Design for Assembly Handbook to Reciprocating Power Saw and Impact Wrench," Report #10, Dept. of Mechanical Engineering, Univ. of Massachusetts, Amherst, Aug. 1980.

16. Karger, W., and Bayha, F.H., *Engineered Work Measurement*, Industrial Press, New York, 1966.

17. Raphael, D.L., "A Study of Positioning Movements," MTM Association of Standards and Research, Fairlawn, N. J., Research Report 109, 1957.

18. Quick, Joseph H., *Work Factor Time Standards*, McGraw-Hill, New York, 1962.

19. Raphael, D. L., "A Study of Arm Movements Involving Weight," Research Report #108, MTM Association of Standards and Research, Fairlawn, N.J., 1957.

Product Design for High-Speed Automatic Assembly and Robot Assembly

8.1 INTRODUCTION

Although design for assembly is an important consideration for manually assembled products and can reap enormous benefits, it is a vital consideration when a product is to be assembled automatically. The simple example shown in Fig. 8.1 serves to illustrate this. The slightly asymmetrical screwed part would not present significant problems in manual handling and insertion whereas, for automatic handling, an expensive vision system would be needed to recognize its orientation. If the part were made symmetrical, automatic handling would be simple. For economic automatic assembly, therefore, careful consideration of product structure and component part design is essential. In fact, it can be said that one of the advantages of introducing automation in the assembly of a product is that it forces a reconsideration of its design, thus reaping not only the benefits of automation but also those of improved product design. Not surprisingly, the savings resulting from product redesign will often outweigh those resulting from automation.

The example of the part in Fig. 8.1 illustrates a further point. The principal problems in applying automation usually involve the automatic handling of the parts rather than their insertion into the assembly. To quote an individual experienced in the subject of automatic assembly, "if a part can be handled automatically, then it can usually be assembled automatically." This means

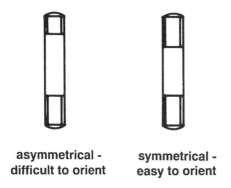

asymmetrical -
difficult to orient

symmetrical -
easy to orient

Fig. 8.1 Design change to simplify automatic feeding and orienting.

that, when we consider design for automation, we will be paying close attention to the design of the parts for ease of automatic feeding and orienting.

In considering manual assembly, we were concerned with prediction of the time taken to accomplish the various tasks such as grasp, orient, insert, and fasten. Then, from a knowledge of the assembly worker's labor rate, we could estimate the cost of assembly. In automatic assembly, the time taken to complete an assembly does not control the assembly cost. Rather it is the rate at which the assembly machine or system cycles because, if everything works properly, a complete assembly is produced at the end of each cycle. Then, if the total rate (cost per unit time) for the machine or system and all the operators is known, the assembly cost can be calculated after allowances are made for downtime. Thus, we shall be concerned mainly with the cost of all the equipment, the number of operators and technicians, and the assembly rate at which the system is designed to operate. However, so that we can identify problems associated with particular parts, we shall need to apportion the cost of product assembly between the individual parts and, for each part, we shall need to know the cost of feeding and orienting and the cost of automatic insertion.

In the following, we shall look first at product design for high-speed automatic assembly using special-purpose equipment, and then we shall consider product design for robot assembly (i.e., using general-purpose equipment).

8.2 DESIGN OF PARTS FOR HIGH-SPEED FEEDING AND ORIENTING

The cost of feeding and orienting parts will depend on the cost of the equipment required and on the time interval between delivery of successive parts. The time between delivery of parts is the reciprocal of the delivery rate and will be nominally equal to the cycle time of the machine or system. If we denote the required delivery or feed rate F_r (parts/min) then, the cost of feeding each part C_f will be given by

$$C_f = \frac{60}{F_r} R_f \text{ cents} \tag{8.1}$$

where R_f is the cost (cents/sec) of using the feeding equipment.

Using a simple payback method for estimation of the feeding equipment rate R_f, this is given by

$$R_f = \frac{C_F E_0}{5760 \, P_b S_n} \text{ cents/sec} \tag{8.2}$$

Where C_F is the feeder cost in dollars, E_0 the equipment factory overhead ratio, P_b the payback period in months, and S_n the number of shifts worked per day.

For example, if we assume that a standard vibratory-bowl feeder costs 5k\$ after installation and debugging, that the payback period is 30 months with two shifts working, and that the factory equipment overheads are 100% ($E_0 = 2$), we get

$$R_f = \frac{5000 \times 2}{5760 \times 30 \times 2}$$

$$= 0.03 \text{ cents/sec}$$

In other words, it would cost 0.03 cents to use the equipment for 1 sec. Suppose that we take this figure as the rate for a "standard" feeder and we assign a relative cost factor C_r to any feeder under consideration; then Eq. (8.1) becomes

$$C_f = 0.03 \frac{60}{F_r} C_r \tag{8.3}$$

Thus, we see that the feeding cost per part is inversely proportional to the required feed rate and proportional to the feeder cost.

To describe these results in simple terms, we can say that, for otherwise identical conditions, it would cost twice as much to feed each part to a machine with a 6-sec cycle compared with the cost for a machine with a 3-sec cycle. This illustrates why it is difficult to justify feeding equipment for assembly systems with long cycle times.

The second result can be simply stated: for otherwise constant conditions, it would cost twice as much to feed a part using a feeder costing 10k\$ compared to a feeder costing 5k\$.

If the feeding cost given by Eq. (8.3) is plotted against the required feed rate F_r on logarithmic scales, a linear relationship results, as shown in Fig.

8.2. However, it appears that the faster the parts are required, the lower the feeding cost. This is true only as long as there is no limit on the speed at which a feeder can operate. Of course, there is always an upper limit to the feed rate obtainable from a particular feeder. We shall denote this maximum feed rate by F_m and consider the factors that affect its magnitude. Before doing so, however, let's look at its effect through an example.

Suppose the maximum feed rate from our feeder is 10 parts/min. Then, if parts are required at a rate of 5 parts/min, the feeder can simply be operated more slowly, involving an increased feeding cost, as given by Eq. (8.3) and illustrated in Fig. 8.2. However, suppose parts are required at a rate of 20 parts/min. In this case, two feeders could be used, each delivering parts at a rate of 10 parts/min. However, the feeding cost per part using two feeders to give twice the maximum feed rate will be the same as one feeder delivering parts at its maximum feed rate. In other words, if the required feed rate is greater than the maximum feed rate obtainable from one feeder, the feeding cost becomes constant and equal to the cost of feeding when the feeder is operating at its maximum rate. This is shown in Fig. 8.2 by the horizontal line. In reality, this line will be saw-toothed, but it can reasonably be assumed that irregularities can be smoothed by spending a little more on feeders to improve their performance when necessary.

From this discussion, we can say that Eq. (8.3) holds true only when the required feed rate F_r is less than the maximum feed rate F_m and, when this is not the case, the feeding cost is given by

$$C_f = 0.03\frac{60}{F_m}C_r \tag{8.4}$$

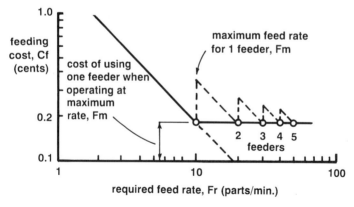

Fig. 8.2 Effect of required feed rate on feeding cost.

Now, the maximum feed rate F_m is given by

$$F_m = 1500\frac{E}{\ell} \text{ parts/sec} \qquad (8.5)$$

where E is the orienting efficiency for the part and ℓ (mm) is its overall dimension in the direction of feeding and where it is assumed that the feed speed is 25 mm/s.

To illustrate the meaning of the orienting efficiency E, we can consider the feeding of dies (cubes with faces numbered 1–6). Suppose that, if no orientation is needed, the dies can be delivered at a rate of 1 per sec from a vibratory-bowl feeder. However, if only those dies with the 6 side uppermost were of interest, a vision system could be employed to detect all other orientations, and a solenoid-operated pusher could be used to reject them. In this case, the delivery rate would fall to an average of one die every 6 sec or a feed rate of 1/6 per sec. If the factor 1/6 is defined as the orienting efficiency, it can be seen that the maximum feed rate is proportional to the orienting efficiency [Eq. (8.5)].

Now, let's suppose that our dies were doubled in size and that the feed speed or conveying velocity on the feeder track was unaffected. It would then take twice as long to deliver each die. In other words, the maximum feed rate is inversely proportional to the length of the part in the feeding direction [Eq. (8.5)].

Equation (8.4) shows that, when $F_r > F_m$, the feeding cost per part is inversely proportional to F_m. It follows that, under these circumstances, the cost of feeding is inversely proportional to the orienting efficiency and proportional to the length of the part in the feeding direction.

This latter relationship illustrates why automatic feeding and orienting methods are applicable only to "small" parts. In practice, this means that parts larger than about 3 in. in their major dimension cannot usually be fed economically.

The designer, when considering the design of a part and its feeding cost, knows the required feed rate and the dimensions of the part. Thus, F_r and ℓ are known. The remaining two parameters that affect feeding cost, namely, the orienting efficiency E and the relative feeder cost C_r, depend on the part symmetry and the types of features that define its orientation. A classification system for part symmetry and features has been developed [1] and, for each part classification, the average magnitudes of E and C_r have been determined [2]. A portion of this system is presented in Figs. 8.3–8.5. Figure 8.3 shows how parts are categorized into basic types, either rotational or nonrotational. For rotational parts, their cylindrical envelopes are classified as disks, short cylinders, or long cylinders. In the case of nonrotational parts, the subcategories are flat, long, or cubic, depending on the dimensions of the sides of the rectangular envelope.

Figure 8.3 gives the first digit of a three-digit shape code. Figure 8.4 shows how the second and third digits are determined for a selection of rotational parts (first digit 0, 1, or 2) and gives the corresponding values of the orienting efficiency E and the relative feeder cost C_r. Similarly, Fig. 8.5 shows how the second and third digits are determined for a selection of nonrotational parts (first digit 6, 7, or 8). The geometrical classification system was originally devised by Boothroyd and Ho [1] as a means of cataloging solutions to feeding problems.

8.3 EXAMPLE

Suppose the part shown in Fig. 8.6 is to be delivered to an automatic assembly station working at a 5-sec cycle. We will now use the classification system and data base to determine the feeding cost, and we will assume that the cost of delivering simple parts at 1/sec using our "standard" feeder is 0.03 cents/part.

First, we must determine the classification code for our part. Figure 8.6 shows that the rectangular envelope for the part has dimensions $A = 30$, $B = 20$, and $C = 15$ mm.

Rotational (1)	Discs $L/D < 0.8$ (2)	0
	Short Cylinders $0.8 \leq L/D \leq 1.5$ (2)	1
	Long Cylinders $L/D > 1.5$ (2)	2
Non-Rotational	Flat $A/B \leq 3$ $A/C > 4$ (3)	6
	Long $A/B > 3$ (3)	7
	Cubic $A/B \leq 3$ $A/C \leq 4$ (3)	8

Fig. 8.3 First digit of geometrical classification of parts for automatic handling (from Ref. 2).

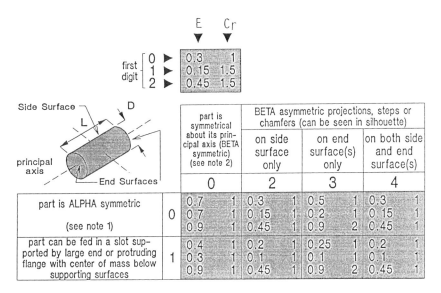

Fig. 8.4 Second and third digits of geometrical classification for some rotational parts (from Ref. 2).

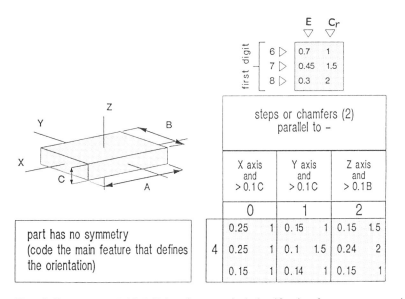

Fig. 8.5 Second and third digits of geometrical classification for some nonrotational parts (from Ref. 2).

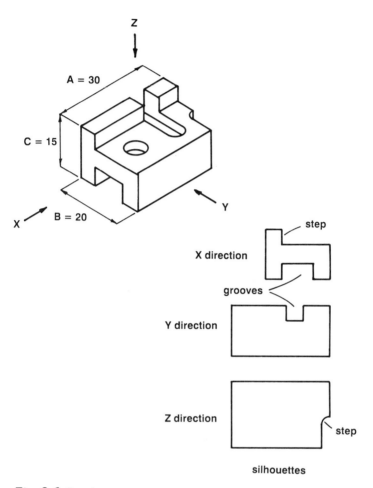

Fig. 8.6 Sample part.

Thus, $A/B = 1.5$ and $A/C = 2$. Referring to Fig. 8.3, we see that, since A/B is less than 3 and A/C is less than 4, the part is categorized as cubic nonrotational and is assigned a first digit of 8. Turning to Fig. 8.5, which provides a selection of data for nonrotational parts, we first determine that our example part has no rotational symmetry about any of its axes. Also, we must decide whether the orientation of the part can be determined by one main feature. Looking at the silhouette of the part in the X direction, we see a step or projection in the basic rectangular shape, and we realize that this feature alone can always be used to determine the orientation of the part. This means that if the silhouette in the X direction is oriented as shown in Fig. 8.6, the part can be in only one orientation and, therefore, the second digit of the classification is 4. However, either the groove apparent in the view in the Y direction and the step seen in the view in the Z direction can also be used to determine the orientation of the part. The procedure now is to select the feature giving the smallest third-classification digit; in this case, it is the step seen in the X direction. Thus, the third digit is 0, giving a three-digit code of 840 and corresponding values of orienting efficiency $E = 0.15$ and relative feeder cost $C_r = 1$.

Since the longest part dimension ℓ is 30 mm and the orienting efficiency E is 0.15, Eq. (8.5) gives the maximum feed rate obtainable from one feeder; thus,

$$F_m = 1500E/\ell$$

$$= 1500 \times 0.15/30$$

$$= 7.5 \text{ parts/min.}$$

Now, from the cycle time of 5 sec, the required feed rate F_r is 12 parts/min, which is slightly higher than F_m. Therefore, since $F_r > F_m$, we use Eq. (8.4) and, since $C_r = 1$, we get a feeding cost of

$$C_f = 0.03 \frac{60}{F_m} C_r$$

$$= 0.03 \frac{60}{7.5} 1$$

$$= 0.24 \text{ cents}$$

8.4 ADDITIONAL FEEDING DIFFICULTIES

In addition to the problems of using the geometric features of the part to orient it automatically, other part characteristics can make feeding particularly difficult. For example, if the edges of the parts are thin, shingling or overlapping can occur during feeding, which leads to problems with the orienting devices on the vibratory-bowl feeder track (Fig. 8.7).

Many other features can affect the difficulty of feeding the part automatically and can lead to considerable increases in the cost of developing the automatic feeding device. These features can also be classified as shown in Fig. 8.8 where, for each combination of features, an approximate additional relative feeder cost is given that should be taken into account in estimating the cost of automatic feeding.

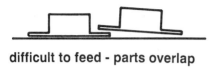

difficult to feed - parts overlap

easy to feed

Fig. 8.7 Parts that shingle or overlap on the feeder track.

				parts will not tangle or nest			
				not light		light	
				not sticky	sticky	not sticky	sticky
				0	1	2	3
parts do not tend to overlap during feeding	not delicate	non-flexible	0	0	1	2	3
		flexible	1	2	3	4	5
	delicate	non-flexible	2	1	2	3	4
		flexible	3	3	4	5	6

Fig. 8.8 Additional relative feeder costs for a selection of feeding difficulties (from Ref. 2).

8.5 HIGH-SPEED AUTOMATIC INSERTION

If a part can be sorted from bulk and delivered to a convenient location correctly oriented, a special-purpose mechanism or workhead can usually be designed that will place it in the assembly. Such workheads can generally be designed to operate on a cycle as short as 1 sec. Thus, for assembly machines operating on cycles greater than 1 sec, the automatic insertion cost C_i will be given by

$$C_i = \frac{60}{F_r} R_i \qquad (8.6)$$

where F_r is the required assembly rate (or feed rate) of parts and R_i is the cost (cents/sec) of using the automatic workhead.

Again, when a simple payback method for estimation of the equipment rate R_i is used, this is given by

$$R_i = \frac{W_c E_0}{5760 P_b S_n} \text{ cents/sec} \qquad (8.7)$$

where W_c is the workhead cost in dollars, E_0 the equipment factory overhead ratio, P_b the payback period in months, and S_n the number of shifts worked per day.

If we assume that a standard workhead costs 10k\$ after installation and debugging, that the payback period is 30 months with two shifts working, and that the factory equipment overheads are 100% ($E_0 = 2$), we get

$$R_i = \frac{10{,}000 \times 2}{5760 \times 30 \times 2}$$

$$= 0.06 \text{ cents/sec}$$

In other words, it would cost 0.06 cents to use the equipment for 1 sec. If we take this figure as the rate for a "standard" workhead and we assign a relative cost factor W_c to any workhead under consideration, Eq. (8.6) becomes

$$C_i = 0.06 \frac{60}{F_r} W_c \qquad (8.8)$$

Thus, the insertion cost is inversely proportional to the required assembly rate and proportional to the workhead cost.

The designer considering the design of a part, knows the required assembly rate F_r. For presentation of relative workhead costs, a classification system for automatic insertion similar to that for manual insertion has been devised [2]. A portion of this system is shown in Fig. 8.9. It can be seen that this classification system is similar to that for manual insertion of parts, except that the first digit is determined by the insertion direction rather than by obstructed access or restricted vision.

8.6 EXAMPLE

If the part shown in Fig. 8.6 is to be inserted horizontally into the assembly in the direction of arrow Y and it is not easy to align and position and is not secured on insertion, the automatic insertion code is 12, giving a relative workhead cost of 1.6.

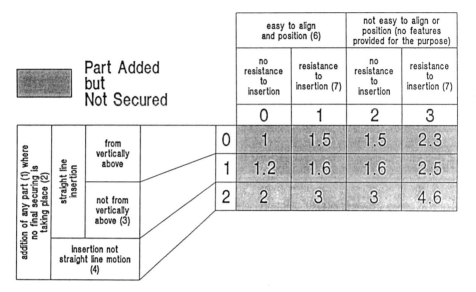

Fig. 8.9 Relative workhead costs W_c for a selection of automatic insertion situations (from Ref. 2).

For a cycle time of 5 sec the assembly rate F_r is 12 parts/min, and Eq. (8.8) gives an insertion cost of

$$C_i = 0.06 \frac{60}{F_r} W_c$$

$$= 0.06 \frac{60}{12} 1.6$$

$$= 0.48 \text{ cents}$$

Thus, the total handling and insertion cost C_t for this part is

$$C_t = C_f + C_i$$

$$= 0.24 + 0.48$$

$$= 0.72 \text{ cents}$$

8.7 ANALYSIS OF AN ASSEMBLY

To facilitate the analysis of a complete assembly, a worksheet similar to that used for manual assembly analysis can be employed. Figure 8.10 shows the exploded view of a simple assembly before and after redesign, which is to be

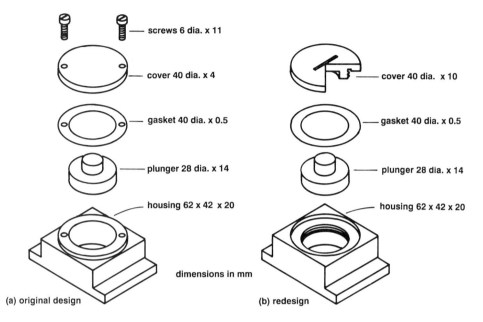

Fig. 8.10 Simple assembly.

assembled at a rate of 9.6/min, and Fig. 8.11 presents the completed work-sheets for automatic assembly analysis. The cost of assembly has been reduced from 16.56 to 2.65 cents.

8.8 GENERAL RULES FOR PRODUCT DESIGN FOR AUTOMATION

The most obvious way to facilitate the assembly process at the design stage is to reduce the number of different parts to a minimum. This subject was covered in the preceding chapter which dealt with manual assembly, where it was emphasized that simplification of the product structure can lead to substantial savings in assembly and parts costs. When considering product design for automation, it is even more important to consider reduction in the number of separate parts. For example, the elimination of a part can eliminate a complete station or an assembly machine, including the parts feeder, the special workhead, and the associated portion of the transfer device. Hence, the reduction in necessary investment can be substantial when a product structure is simplified.

Apart from product simplification, automation can be facilitated by the introduction of guides and chamfers that directly facilitate assembly. Examples of this are given by Baldwin [3] and Tipping [4] in Figs. 8.12 and 8.13. In

(a) original design — HIGH-SPEED AUTOMATIC ASSEMBLY — Name of Assembly: VALVE

ID	RP	HC	OE	CR	FM	DF	CF	IC	WC	DI	CI	CA	NM	Name of Part, Sub-assembly or Operation
1	1	83100	0.20	1	4.8	12.4	0.40	00	1.0	6.3	0.29	0.69	1	housing
2	1	02000	0.40	1	21.4	6.3	0.20	02	1.5	9.4	0.43	0.63	1	plunger
3	1	00840	.*	*	***.*	**.*	*.**	-manual ass'y required-				7.13	0	gasket
4	1	00800	.*	*	***.*	**.*	*.**	-manual ass'y required-				6.67	1	cover
5	2	21000	0.90	1	122.7	6.3	0.20	39	1.8	11.3	0.52	1.44	0	screw

Column key: RP = No. of Repeats; HC = Handling Code; OE = Orientation Efficiency; CR = Relative Feeder Cost; FM = Maximum Feed Rate (parts/min.); DF = Handling Difficulty; CF = Handling Cost (cents); IC = Insertion Code; WC = Relative Workhead Cost; DI = Insertion Difficulty; CI = Insertion Cost (cents); CA = Total Cost (cents); NM = Figure for min. parts.

(a) original design

(b) redesign — HIGH-SPEED AUTOMATIC ASSEMBLY — Name of Assembly: NEWVALVE

ID	RP	HC	OE	CR	FM	DF	CF	IC	WC	DI	CI	CA	NM	Name of Part, Sub-assembly or Operation
1	1	83100	0.20	1	4.8	12.4	0.40	00	1.0	6.3	0.29	0.69	1	housing
2	1	02000	0.40	1	21.4	6.3	0.20	02	1.5	9.4	0.43	0.63	1	plunger
3	1	00040	0.70	3	26.3	18.8	0.61	00	1.0	6.3	0.29	0.90	0	gasket
4	1	02000	0.40	1	15.0	6.3	0.20	38	0.8	5.0	0.23	0.43	1	cover

(b) redesign

Fig. 8.11 Completed worksheets for high-speed automatic assembly analysis of the assemblies in Fig. 8.10.

both examples, sharp corners are removed so that the part to be assembled is guided into its correct position during assembly and requires less control by the placement device. This measure can even eliminate the need for a placement device.

Further examples in this category can be found in the types of screws used in automatic assembly. Those screws that tend to centralize themselves in the

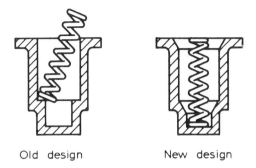

Old design   New design

Fig. 8.12 Redesign of part for ease of assembly (from Ref. 3).

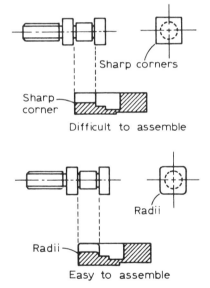

Fig. 8.13 Redesign to assist assembly (from Ref. 4).

hole give the best results in automatic assembly. Tipping [4] summarizes and grades the designs of screw points available as follows (Fig. 8.14):

1. Rolled thread point: very poor location; does not centralize without positive control on the outside diameter of the screws
2. Header point: only slightly better than point 1 if of correct shape

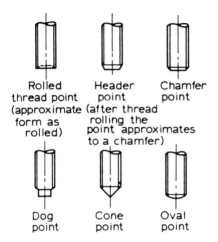

Fig. 8.14 Various forms of screw points (from Ref. 4).

3. Chamfer point: reasonable to locate
4. Dog point: reasonable to locate
5. Cone point: very good to locate
6. Oval point: very good to locate

Tipping recommends that only the cone- and oval-point screws be used in automatic assembly.

Another factor to be considered in design is the difficulty of assembly from directions other than directly above. The aim of the designer should be to allow for assembly in sandwich or layer fashion, with each part placed on top of the previous one. The biggest advantage of this method is that gravity is useful in the feeding and placing of parts. It is also desirable to have work-heads and feeding devices above the assembly station, where they will be accessible in the event of a fault due to the feeding of a defective part. Assembly from above may also assist in the problem of keeping parts in their correct positions during the machine index period, when dynamic forces in the horizontal plane might tend to displace them. In this case, with proper product design, where the parts are self-locating, the force due to gravity should be sufficient to hold the part until it is fastened or secured.

If assembly from above is not possible, it is probably wise to divide the assembly into subassemblies. For example, an exploded view of a British power plug is shown in Fig. 8.15 and, in the assembly of this product, it would be relatively difficult to position and drive the two cord-grip screws

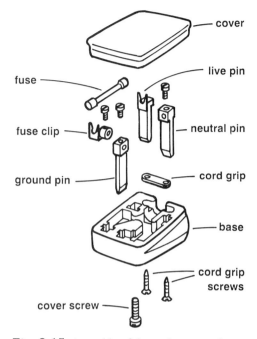

Fig. 8.15 Assembly of three-pin power plug.

from below. The remainder of the assembly (apart from the main holding screw) can be conveniently built into the base from above. In this example, the two screws, the cord grip, and the plug base could be treated as a subassembly dealt with prior to the main assembly machine.

It is always necessary in automatic assembly to have a base part on which the assembly can be built. This base part must have features that make it suitable for quick and accurate location on the work carrier. Figure 8.16a shows a base part for which a suitable work carrier would be difficult to design. In this case, if a force is applied at A, the part will rotate unless adequate clamping is provided. One method of ensuring that a base part is stable is to arrange that its center of gravity is contained within flat horizontal surfaces. For example, a small ledge machined into the part allows a simple and efficient work carrier to be designed (Fig. 8.16b).

Location of the base part in the horizontal plane is often achieved by dowel pins mounted in the work carrier. To simplify the assembly of the base part onto the work carrier, the dowel pins can be tapered to provide guidance, as in the example shown in Fig. 8.17.

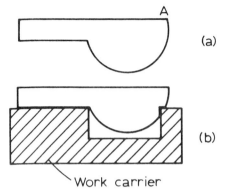

Fig. 8.16 Design of base part for mounting on work carrier.

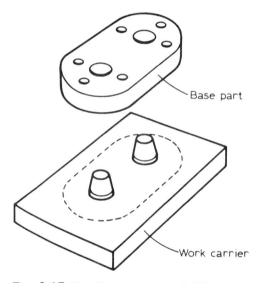

Fig. 8.17 Use of tapered pegs to facilitate assembly.

8.9 DESIGN OF PARTS FOR FEEDING
AND ORIENTING

Many types of parts feeders are used in automatic assembly, and some of them have been studied in chapters 3 and 4. Most feeders are suitable for feeding only a very limited range of part shapes and are not generally relevant to discussions of the design of parts for feeding and orienting. The most versatile parts feeder is the vibratory-bowl feeder, and this section deals mainly with the aspects of the design of parts that will facilitate feeding and orienting in this type of feeder. Many of the points made, however, apply equally to other feeding devices. Three basic design principles can be enumerated:

1. Avoid designing parts that will tangle, nest, or shingle.
2. Make the parts symmetrical.
3. If parts cannot be made symmetrical, avoid slight asymmetry or asymmetry resulting from small or nongeometrical features.

It can be almost impossible to separate, orient, and feed automatically parts that tend to tangle or nest when stored in bulk. Often, a small nonfunctional change in design will prevent this occurrence; some simple examples of this are illustrated in Fig. 8.18.

While the asymmetrical feature of a part might be exaggerated to facilitate orientation, an alternative approach is to deliberately add asymmetrical features for the purpose of orienting. The latter approach is more common,

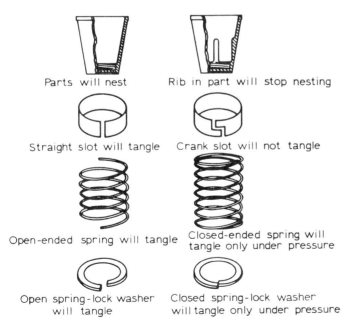

Parts will nest   Rib in part will stop nesting

Straight slot will tangle   Crank slot will not tangle

Open-ended spring will tangle   Closed-ended spring will tangle only under pressure

Open spring-lock washer will tangle   Closed spring-lock washer will tangle only under pressure

Fig. 8.18 Examples of redesign to prevent nesting or tangling (from Ref. 5).

and some examples, given by Iredale, [5] are reproduced in Fig. 8.19. In each case, the features that require alignments are difficult to utilize in an orienting device, and so corresponding external features are added deliberately.

It will be noted that, in the portions of the coding systems shown in Figs. 8.4 and 8.5, those parts with a high degree of symmetry all have codes representing parts that are easy to handle. There are, however, a wide range of codes representing parts that will probably be difficult to handle automatically; designers will require assistance with the problems these parts create.

Figure 8.20a shows a part that would be difficult to handle, and Fig. 8.20b shows the redesigned part, which could be fed and oriented in a vibratory-bowl feeder at a high rate. The subtle change in design would not be obvious to the designer without the use of the coding system. Without it, it might not have occurred to the designer that the original part was difficult to handle automatically.

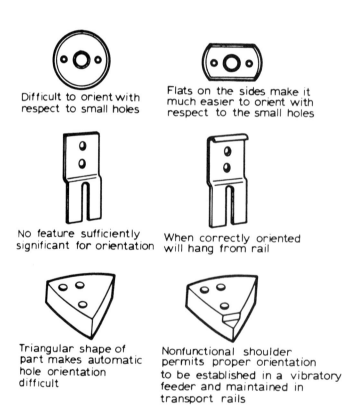

Difficult to orient with respect to small holes

Flats on the sides make it much easier to orient with respect to the small holes

No feature sufficiently significant for orientation

When correctly oriented will hang from rail

Triangular shape of part makes automatic hole orientation difficult

Nonfunctional shoulder permits proper orientation to be established in a vibratory feeder and maintained in transport rails

Fig. 8.19 Provision of asymmetrical features to assist in orientation (from Ref. 5).

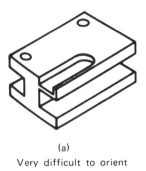 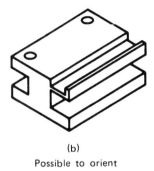

(a)   (b)
Very difficult to orient   Possible to orient

Fig. 8.20 Less obvious example of a design change to simplify feeding and orienting.

It should be pointed out that, although the discussion above dealt specifically with automatic handling, parts that are easy to handle automatically are also easy to handle manually. A reduction in the time taken for an assembly worker to recognize the orientation of a part and then reorient it results in considerable cost savings.

Clearly, with some parts, it will not be possible to make design changes that will enable them to be handled automatically: for example, very small parts or complicated shapes formed from thin strips are difficult to handle in an automatic environment. In these cases, it is sometimes possible to manufacture the parts on the assembly machine or to separate them from the strip at the moment of assembly. Operations such as spring winding or blanking out thin sections have been successfully introduced on assembly machines in the past.

8.10 SUMMARY OF DESIGN RULES FOR HIGH-SPEED AUTOMATIC ASSEMBLY

The various points made in this discussion of parts and product design for automatic assembly are summarized below in the form of simple rules for the designer.

8.10.1 Rules for Product Design

1. Minimize the number of parts.
2. Ensure that the product has a suitable base part on which to build the assembly.
3. Ensure that the base part has features that will enable it to be readily located in a stable position in the horizontal plane.
4. If possible, design the product so that it can be built up in layer fashion, each part being assembled from above and positively located

so that there is no tendency for it to move under the action of horizontal forces during the machine index period.

5. Try to facilitate assembly by providing chamfers or tapers that will help to guide and position the parts in the correct position.
6. Avoid expensive and time-consuming fastening operations, such as screwing, soldering, and so on.

8.10.2 Rules for the Design of Parts

1. Avoid projections, holes, or slots that will cause tangling with identical parts when placed in bulk in the feeder. This may be achieved by arranging that the holes or slots are smaller than the projections.
2. Attempt to make the parts symmetrical to avoid the need for extra orienting devices and the corresponding loss in feeder efficiency.
3. If symmetry cannot be achieved, exaggerate asymmetrical features to facilitate orienting or, alternatively, provide corresponding asymmetrical features that can be used to orient the parts.

8.11 PRODUCT DESIGN FOR ROBOT ASSEMBLY

As with product design for high-speed automatic assembly, one objective here is to provide the designer with a means of estimating the cost of assembling the product—but, in this case, using robots. However, several important design aspects will be affected by the choice of robot assembly system, a choice that, in turn, is affected by various production parameters such as production volume and the number of parts in the assembly. In Chapter 6, it was seen that three representative types of robot assembly systems can be considered, namely:

1. Single-station with one robot arm
2. Single-station with two robot arms
3. Multistation with robots, special-purpose workheads, and manual assembly stations as appropriate

For a single-station system, those parts that require manual handling and assembly and that must be inserted during the assembly cycle present special problems. For reasons of safety, it is usually necessary to transfer the assembly to a location or fixture outside the working environment of the robot. This can be accomplished by having the robot place the assembly on a transfer device that carries the assembly to the manual station. After the manual operation has been completed, the assembly can be returned in a similar manner to within reach of the robot.

The use of special-purpose workheads for insertion or securing operations presents problems similar to those for manual assembly operations. Two different situations can be encountered. In the first, the robot inserts or places the part without securing it immediately. This operation is followed by transfer of the assembly to an external workstation where the securing operation is carried out; a heavy press fit is an example. In the second situation, a special-purpose workhead is engineered to interact directly at the robot work fixture. This might take the form of equipment activated from the sides of, or underneath, the work fixture to carry out soldering, tab bending or twisting operations, spin riveting, etc., while the robot has to place and, if necessary, manipulate the part.

These major problems with single-station systems do not occur with the multistation system, where manual operations or special-purpose workheads can be assigned to individual stations as necessary. This illustrates why it is important to know the type of assembly system likely to be employed when the product is being designed.

In order to determine assembly costs, it is necessary to obtain estimates of the following:

1. The total cost of all the general-purpose equipment used in the system, including the cost of robots and any transfer devices and versatile grippers, all of which can be employed in the assembly of other products if necessary
2. The total cost of all the special-purpose equipment and tooling, including special-purpose workheads, special fixtures, special robot tools or grippers, and special-purpose feeders; and special magazines, pallets, or part trays
3. The average assembly cycle time, that is, the average time to produce a complete product or assembly
4. The cost per assembly of the manual labor involved in machine supervision, loading feeders, magazines, pallets, or part trays, and the performance of any manual assembly tasks.

Classification systems and data bases have been developed for this purpose and are included in the *Product Design for Assembly Handbook* [2]. The information presented in the handbook allows all these estimates to be made and includes one classification and data chart for each of the three basic robot assembly systems. In these charts, insertion or other required operations are classified according to difficulty. For each classification, and depending on the difficulty of the operation, relative cost and time factors are given that can be used to estimate equipment costs and assembly times. These cost and time estimates are obtained by entering data from the appropriate chart onto a worksheet for each part insertion or separate operation.

Figure 8.21 shows a portion of the classification system and data base for a single-station one-arm robot assembly system. This portion of the system applies to the situation in which a part is added to the assembly but is not secured immediately. The selection of the appropriate row (first digit) depends on the direction of insertion, an important factor influencing the choice of robot because the SCARA-type assembly robot like the IBM 7535 (Fig. 5.41) can perform insertions only along the vertical axis. The selection of the appropriate column (second digit) depends on whether the part needs a special gripper and temporary clamping after insertion, and whether it tends to align itself during insertion. All these factors affect either the cost of the tooling required or the time for the insertion operation or both.

When the row and columns have been selected for a particular operation, the figures in the box allow estimates to be made of the robot cost, the gripper or tool cost, and the total time for the operation.

Let's suppose that a part is to be inserted along a horizontal axis, does not require a special gripper, requires temporary clamping, and is easy to align. For this operation, the code would be 12. In this case, the relative robot cost AR is 1.5. This means that, if the basic capital cost of the installed standard four-degree-of-freedom robot (including all controls, sensors, etc., and capable of only vertical insertions) is 60k$, a cost of 90k$ is assumed. This figure

Single-Station One-Arm System

AR	TP
AG	TG

TP - relative affective basic operation time
AR - relative robot cost
AG - relative additional gripper or tool cost
TG - relative time penality for gripper or tool change

part added but not finally secured		part can be gripped & inserted using standard gripper or gripper used for previous part			
		no holding down		part requires temporary holding or clamping	
		self-aligning	not easy to align	self-aligning	not easy to align
		0	1	2	3
using motion along or about the vertical axis	0	1.0 1.0 0 0	1.0 1.07 0 0	1.0 1.0 1.0 0	1.0 1.07 1.0 0
using motion along or about a non-vertical axis	1	1.5 1.0 0 0	1.5 1.07 0 0	1.5 1.0 1.0 0	1.5 1.07 1.0 0
involving motion along or about more than one axis	2	1.5 1.8 0 0	1.5 1.9 0 0	1.5 1.8 1.0 0	1.5 1.9 1.0 0

Fig. 8.21 Portion of classification system and data base for a single-station one-arm robot assembly system (from Ref. 2).

allows for a more sophisticated robot, able to perform operations from directions other than above. In other words, there is a cost penalty of 30k$ for the basic equipment in the system because the "standard robot" cannot perform the operation required.

The value of the relative additional gripper or tool cost is 1.0. Since the part needs temporary clamping, special tooling mounted on the work fixture would be required. Thus, if the standard tooling or gripper costs 5k$, the additional tooling needed would represent a cost penalty of 5k$ in the form of special-purpose equipment.

The value of the relative basic operation time TP is 1.0. In this analysis method, the basis for time estimates is the average time taken by the robot to move approximately 0.5 m, grasp the part, return, and insert the part when the motion is simple and no insertion problems exist. For a typical present-generation robot, this process might take 3 sec. If this figure is used in the present example, this is the basic time for the robot to complete the operation.

Finally, since the relative time penalty for gripper or tool change is zero, no additional time penalty is incurred, and the total operation time is 3 sec. In some cases such as that in which the part to be inserted is not completely oriented by the part-presentation device, a further time penalty must be added. In this case, the robot arm must perform the final orientation with the aid of a simple vision system, and an additional 2–3 sec must be added to the operation time.

In addition to the cost of the robot and the special tools or grippers, the costs of part presentation must be estimated. Before this can be accomplished, it must be decided which part-presentation method will be used for each part. In practice, there are usually only two choices, namely, (1) a special-purpose feeder or (2) a manually loaded magazine, pallet, or part tray.

The costs associated with part presentation can be divided into:

1. Labor costs, which include material handling (loading parts feeders or magazines), system tending (freeing jams in feeders, handling parts trays, etc.), and system changeover costs (changing of work fixture and feeders and magazines and reprogramming of robots)
2. Equipment costs, including the depreciation of feeders, special fixtures, special tooling, and magazines, pallets, or part trays

It can be assumed that the bulk material handling costs (i.e., dumping parts in bulk into feeder hoppers) are negligible compared to the cost of manually loading individual parts one by one into magazines, pallets, or part trays.

There are thus only three significant factors needed to estimate the cost of part presentation:

1. Special-purpose feeders. The cost of a special-purpose feeder, fully tooled and operating on the robot system, is assumed to be a minimum of 5k$. The actual cost of a feeder, for a particular part, can be obtained from the data presented earlier in this chapter, where feeding and orienting costs were considered in detail.

2. Manually loaded magazines. The cost of one set of special magazines, pallets, or part trays for one part type is assumed to be 1k$. For larger parts, this may considerably underestimate the actual cost, and extra allowance should be made.

3. Loading of magazines. The time to hand-load one part into a magazine can be estimated to be the part-handling time, obtained from the data in Chapter 7, plus 1 sec. Alternatively, a typical value of 4 sec may be used.

It can be seen that use of the classification systems and data base allows the total cost of equipment and the cost of any manual assembly work to be estimated, together with the assembly time for each part. These results provide the data necessary to predict assembly costs using each of the three robot assembly systems. A completed analysis is presented in chapter 10, in which the costs of robot assembly, high-speed automatic assembly, and manual assembly are compared.

8.11.1 Summary of Design Rules for Robot Assembly

Many of the rules for product design for manual assembly and high-speed automatic assembly also apply to product design for robot assembly. However, when we weigh the suitability of a proposed design for robot assembly, we should carefully consider the need for any special-purpose equipment such as special grippers or special feeders. The cost of this equipment must be amortized over the total life volume of the product and, for the midrange volumes to which robot assembly might be applied, this can add considerably to the cost of assembly.

The following are some specific rules to follow during product design [2]:

1. Reduce part count; this is a major strategy for reducing assembly, manufacture, and overhead costs, irrespective of the assembly system to be used.

2. Include features such as leads, lips, and chamfers to make parts self-aligning in assembly. Because of the relatively poor repeatability of many robot manipulators compared to dedicated workhead mechanisms, this is a vitally important measure to ensure consistent fault-free part insertions.

3. Ensure that parts that are not secured immediately on insertion are self-locating in the assembly. For multistation robot assembly systems or one-arm single-station systems, this is an essential design rule. Holding down of unsecured parts cannot be carried out by a single robot arm, and so special

fixturing is required, which must be activated by the robot controller. This adds significantly to special-purpose tooling and, hence, assembly costs. With a two-arm single-station system, one arm can, in principle, hold down an unsecured part while the other continues the assembly and fastening processes. In practice, this requires one arm to change end-of-arm tooling to a hold-down device; the system then proceeds with 50% efficiency while one arm remains immobile.

4. Design parts so that they can all be gripped and inserted using the same robot gripper. One major cause of inefficiency with robot assembly systems arises from the need for gripper or tool changes. Even with rapid gripper or tool-change systems, each change in a special gripper and then back to the standard gripper is approximately equal to two assembly operations. Note that the use of screw fasteners always results in the need for tool changes since robot wrists can seldom rotate more than one revolution.

5. Design products so that they can be assembled in layer fashion from directly above (Z-axis assembly). This ensures that the simplest, least costly, and most reliable four-degree-of-freedom robot arms can accomplish the assembly tasks. It also simplifies the design of the special-purpose work fixture.

6. Avoid the need for reorienting the partial assembly or manipulating previously assembled parts. These operations increase the robot assembly cycle time without adding value to the assembly. Moreover, if the partial assembly has to be turned to a different resting aspect during the assembly process, this will usually result in increased work-fixture cost and the need to use a more expensive six-degree-of-freedom robot arm.

7. Design parts that can be easily handled from bulk. To achieve this goal, avoid parts that

nest or tangle in bulk

are flexible

have thin or tapered edges that can overlap or "shingle" when moving along a conveyor or feed track

are delicate or fragile to the extent that recirculation in a feeder can cause damage

are sticky or magnetic so that a force comparable to the weight of the part is required for separation

are abrasive and will wear the surfaces of automatic handling systems

are light so that air resistance will create conveying problems (less than 1.5 N/m^3 or 0.01 lb/in^3.)

8. If parts are to be presented using automatic feeders, ensure that they can be oriented using simple tooling. Follow the rules for ease of part orientation discussed earlier. Note, however, that feeding and orienting at high speed

is seldom necessary in robot assembly and that the main concern is that the features defining part orientation can be easily detected.

9. If parts are to be presented using automatic feeders, ensure that they can be delivered in an orientation from which they can be gripped and inserted without any manipulation. For example, avoid situations in which a part can be fed in only one orientation from which it must be turned over for insertion. This will require a six-degree-of-freedom robot and a special gripper or a special 180-deg-turn delivery track; both solutions lead to unnecessary cost increases.

10. If parts are to be presented in magazines or part trays, ensure that they have a stable resting aspect from which they can be gripped and inserted without any manipulation by the robot. It should be noted that, if the production conditions are appropriate, the use of robots holds advantages over the use of special-purpose workheads and that some design rules can be relaxed. For example, a robot can be programmed to acquire parts presented in an array, such as a pallet or part tray that has been loaded manually, thus avoiding many of the problems arising with automatic feeding from bulk. When making economic comparisons, however, the cost of manual loading of the magazines must be taken into account.

REFERENCES

1. Boothroyd, G., and Ho, C., "Coding System for Small Parts for Automatic Handling," Society of Manufacturing Engineers, Paper ADR 76-13, Assemblex III Conference, Chicago, Oct. 1976.
2. Boothroyd, G., and Dewhurst, P., "Product Design for Assembly Handbook," Boothroyd Dewhurst, Inc., Wakefield, R.I., 1986.
3. Baldwin, S.P., "How to Make Sure of Easy Assembly," *Tool and Manufacturing Engineering,* May, 1966, p. 67.
4. Tipping, W.V., "Component and Product Design for Mechanized Assembly," Conference on Assembly, Fastening and Joining Techniques and Equipment, Production Engineering Research Association, England, 1965.
5. Iredale, R., "Automatic Assembly—Components and Products," *Metalwork Production,* April 8, 1964.

9

Printed-Circuit-Board Assembly

9.1 INTRODUCTION

Printed-circuit-board (*PCB*) assembly involves mainly the insertion and solder-ing of electrical *components* into printed circuit boards. Component insertion is carried out manually or by high-speed, dedicated machinery. Additionally, some manufacturers employ robots to perform insertions, mainly those that must otherwise be performed manually because of the *nonstandard* shapes of the components. For high production volumes, most manufacturers use a com-bination of automatic and manual insertion because odd-shaped or nonstandard components cannot be handled by the automatic insertion machines. It is desir-able, however, to use automatic insertion machines wherever possible since they can operate much faster and with greater reliability than manual workers. For those PCBs manufactured in small batches and where the applications involve severe working environments, such as military applications, assembly is performed entirely by hand.

At the end of this chapter, data and equations are presented that can be used to estimate the cost of component insertion and soldering either by dedicated automatic insertion machines, manual assembly workers, or robots.

9.2 TERMINOLOGY

For those not familiar with the terms used in PCB manufacture, a glossary of the terminology is included at the end of this chapter.

The term *insertion* is used to describe the process of placing a through-hole electrical *component* onto a printed circuit board so that its leads pass through the correct holes in the board or of placing a *surface-mount* component onto the board in the required position. There are three methods of insertion: (1) dedicated automatic insertion machines; (2) manual assembly workers, including semiautomatic insertion; and (3) robots.

Automatic insertion of axial (*VCD*) components involves *preforming* the leads, inserting the component, and cutting and clinching the leads. Preforming and cut-and-clinch are done automatically as part of the insertion cycle and do not add to the cycle time or decrease the rate of insertion. Automatic insertion of *DIPs* does not involve lead forming or cutting.

With manual insertion and semiautomatic insertion, all the operations are performed manually in sequence. Thus, preforming before insertion and cut-and-clinch after insertion add to the total time for the insertion operation.

Robot insertion involves movement of the robot arm to the component, grasping the component, realigning it if necessary, moving it to the correct board location, and inserting it. If the component feeders or presenters cannot completely orient the component, robot insertion will include final realignment and will increase the cycle time. As with manual assembly, preforming and cut-and-clinch are usually done separately.

Robots are generally used only to insert nonstandard components that otherwise would have been inserted manually. Nonstandard, or odd-form, components are those large or small or odd-shaped components that cannot be inserted by special-purpose machinery.

Before introducing the cost analysis for printed-circuit-board assembly, the complete assembly procedure will be described. The following section explains the various steps that can be included in a PCB assembly process; these steps are automatic, manual, and robotic insertion of components.

9.3 ASSEMBLY PROCESS FOR PRINTED-CIRCUIT BOARDS

Figure 9.1 shows all the possible steps in an assembly process for PCBs [1]. The figure includes that portion of PCB manufacture in which component insertions are carried out. Component presentation, repair for faulty insertions, and *touch-up* for faulty soldering are also included. Steps that do not directly involve the addition of components to the board, however, such as board preparation, where boards are given identification codes; board cut, where a

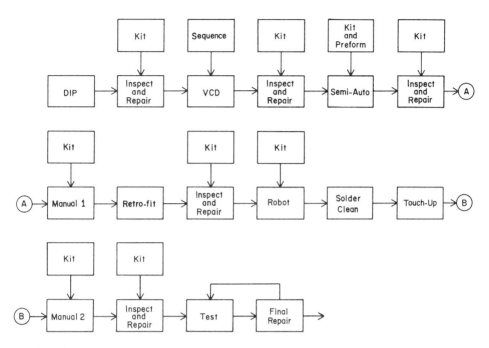

Fig. 9.1 Typical assembly process for printed circuit boards (from Ref. 1).

series of identical boards is cut from a larger panel; and final test, where functional testing of the board is done, are not included. Boards move through the steps as indicated by the horizontal lines on the figure. The vertical lines indicate the flow of components from inventory to the insertion stations.

In order to minimize handling and processing time, small boards are sometimes processed in the form of larger panels which, after the components are inserted and soldered, are separated into the individual boards. Alternatively, several separate boards can be mounted in one fixture to be processed together.

The first block in the assembly process shown in Fig. 9.1 indicates DIP. This refers to the automatic insertion of dual in-line package components (DIPs) and includes all integrated circuit chips and chip sockets. The term *dual in-line package* refers to the two parallel rows of leads projecting from the sides of the package (Fig. 9.2). Typically, DIPs have between 4–40 leads; DIPs with more than 40 leads are infrequently used. The lead span, which is the distance between the two rows of leads, is standardized at 0.3, 0.4, or 0.6 in.

At the DIP station, components are inserted with an automatic DIP inserter (Fig. 9.3). Automatic insertion is carried out at high rates, approximately

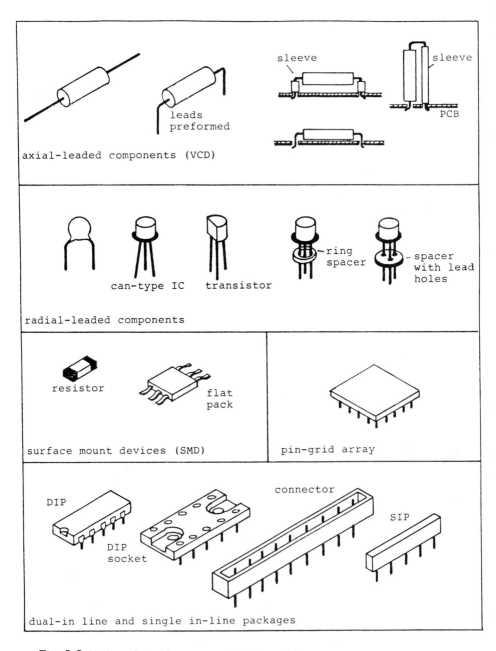

Fig. 9.2 Various electronic components (not to scale).

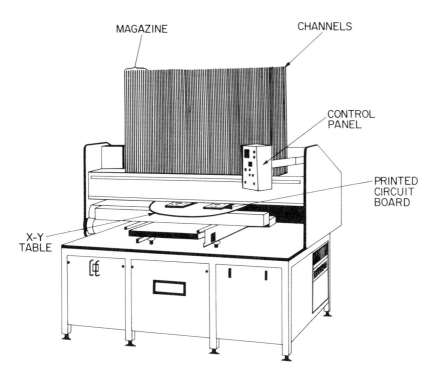

MAGAZINE

CHANNELS

CONTROL PANEL

PRINTED CIRCUIT BOARD

X-Y TABLE

Fig. 9.3 Automatic DIP insertion machine.

2800–4500/hr. or one component every 0.80–1.29 sec. The DIP leads are inserted through predrilled holes in the board and are cut and clinched below the board (Fig. 9.4). The automatic insertion head moves only in the vertical direction while the board is positioned below it on an x-y table, which can also rotate. High-performance DIP insertion machines can insert DIPs having any of the three standard lead spans with no tooling change, but base models can handle only 0.3-in. lead spans and 6–20-lead DIPs. To accommodate 2- and 4-lead DIPs, a special insertion head must be employed at additional cost. To check for electrical *faults*, a component verifier may be employed that will stop the machine if certain preprogrammed electrical characteristics are not met.

Dual in-line package components are purchased from component manufacturers in long tubes called *channels*, in which the components are preoriented and stacked end to end. The channels are loaded onto the DIP inserting machine. Usually, one channel is used for each DIP type on the board but, if

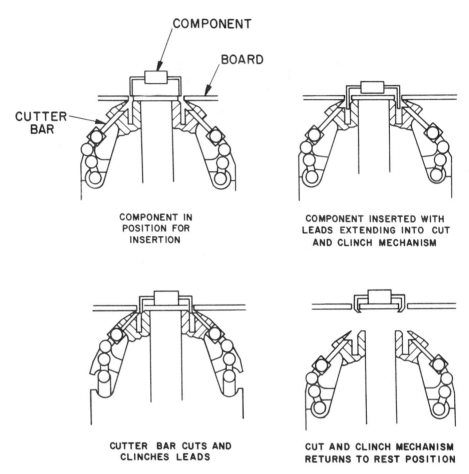

Fig. 9.4 Cut-and-clinch sequence.

large numbers of one type are used, more channels can be assigned to the same component. A *magazine* refers to a group of channels, usually about 15. If a high component mix is required, additional magazines can be added to the machine. The machine size and speed restricts the number of magazines. The insertion cycle time is longer when the channel is farther from the insertion head. Channels can be changed by an operator as the machine is running, which eliminates downtime caused by empty channels.

After DIP insertion, the next block in the assembly process shown in Fig. 9.1 indicates inspect and repair. This is an inspection of the partially assembled board by the inspector who is looking for faults that can be detected visually, such as broken or bent leads or components inserted into the wrong holes. Components are either repaired and reinserted or discarded and replaced. Workers have available at each repair station all the components inserted at the previous insertion station, so that any component may be replaced. Inspect-and-repair stations can follow each insertion station. However, workers can never detect all the faults, and those that go undetected, will inevitably have to be detected and corrected later in the manufacturing process.

The second insertion station in the assembly process is VCD insertion. This refers to the automatic insertion of *axial-lead components* (Fig. 9.2), also called VCDs (variable center distance components). These include resistors, capacitors, and diodes within the size limitations of the insertion head. Axial-lead components must usually have their leads bent at right angles prior to insertion, and the final lead span (called *center distance*) is variable.

At this second station, axial-lead components are inserted with an automatic axial-lead inserter (Fig. 9.5). Axial-lead components are inserted at rates of approximately 9,500 to 32,000 components/hr or one component every 0.11–0.38 sec. Rates of 32,000/hr can be achieved only with a dual-head insertion

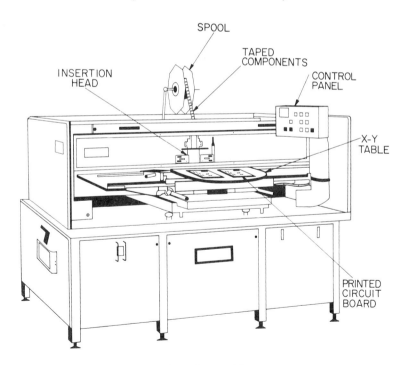

Fig. 9.5 Automatic axial-lead insertion machine.

machine, a machine that inserts components into two identical boards simultaneously. Obviously, single-head axial-lead insertion machines work at half the rate of dual-head machines. The rates are higher than those for automatic DIP insertion because components are fed on a *spool* in the correct sequence for insertion (Fig. 9.6) and do not have to be moved as far to reach the insertion head. Spools can hold large numbers of components and only have to be changed infrequently.

Automatic axial-lead insertion proceeds as follows:

1. Components are stored on a spool in the correct order for insertion (this is accomplished by the automatic component sequencer described below).
2. The spool is loaded manually onto the axial-lead inserter,
3. During the automatic insertion cycle, the leads are automatically cut to remove the component from the tape; then they are bent at right angles

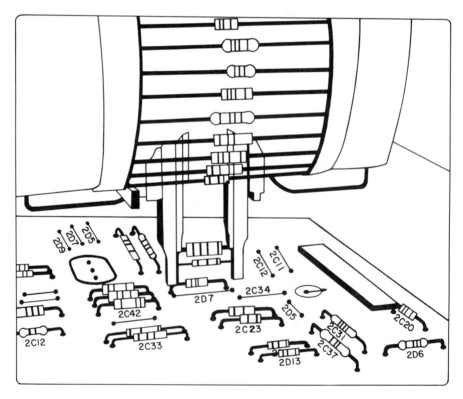

Fig. 9.6 Axial-lead components on tape at insertion head.

to the correct center distance, and the component is positioned with the leads passing through the board.

4.   Finally, the leads are cut and clinched below the board (Fig. 9.4).

As indicated in Fig. 9.1, axial-lead components are first mounted on a spool using a component *sequencer*. This machine (Fig. 9.7) arranges the components on a tape spool in the correct sequence for insertion. Axial-lead components are purchased on spools, with one component type on each spool. Spools are loaded manually onto the sequencer for every type of component on the board and, from these, a master spool is created.

A component sequencer can handle components at a rate of approximately 10,000–25,000/hr or one component every 0.14–0.36 sec. Component leads are automatically cut to remove the components from their individual tapes in the correct order for insertion. Components are then retaped and rolled onto the master spool. Master spools are made off-line and in advance of the component insertion process.

A *module* for a sequencer is a group of dispensing spools, usually about 20 in number. If a larger component mix is needed, additional modules can be

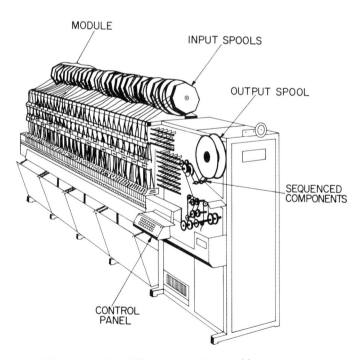

Fig. 9.7 Automatic axial-lead sequencing machine.

added to the sequencer up to a maximum of about 240 dispensing spools, with the limit being imposed by the physical size of the machine. To check for faulty components or a component out of order, a component verifier can be added to the sequencer.

These last two processes, automatic axial-lead component insertion and sequencing, may be combined and carried out on one machine. Such a machine, about twice as large as a conventional automatic insertion machine, eliminates the need for *kitting* and reduces setup time by bringing inventory to the assembly line. Since components are stored at the assembly line, different batches can be run, both large and small, by simply writing a new program for each batch.

In addition to these automatic insertion processes, PCB manufacturers may include stations for the automatic insertion of *radial-lead components* and single in-line package (*SIP*) components (Fig. 9.2). Automatic insertion machines for radial-lead components are similar to automatic axial-lead component insertion machines, and automatic SIP inserters are similar to automatic DIP inserters. Some DIP insertion machines have as an option the ability to insert SIP components using an additional insertion head. Also, the first three stations in Fig. 9.1 may each include duplicate machines if the number of components per board is high.

The next block in the PCB assembly process shown in Fig. 9.1 indicates semiauto. This refers to semiautomatic component insertion, i.e., machine-assisted manual insertion. Inserted at this station are all DIP and axial-lead components that cannot be machine-inserted because of either their size or their location on the board. Also inserted are radial-lead components, SIPs, and some connectors. Wherever possible in high-volume assembly, semiautomatic insertion is used instead of manual insertion since it can reduce insertion times by 80%.

A semiautomatic insertion machine (Fig. 9.8) automatically presents the correct component to the operator and indicates, by use of a light beam, the correct location and orientation for the component on the board. The component is then inserted by hand, with the leads sometimes being automatically cut and clinched. Components are inserted in this way at rates of approximately one component every 5 sec. All components must have their leads preformed to the correct dimensions for insertion prior to presentation to the operator. Typically, components are stored in a rotating tray. After receiving a signal from the operator, the tray rotates so that only the section containing correct components is presented. The light beam (which can be located either above or below the board) illuminates the holes for the insertion and uses a symbol to indicate component polarity.

There are two manual assembly stations, Manual 1 and Manual 2, shown in the PCB assembly process in Fig. 9.1. One station is before, and the other

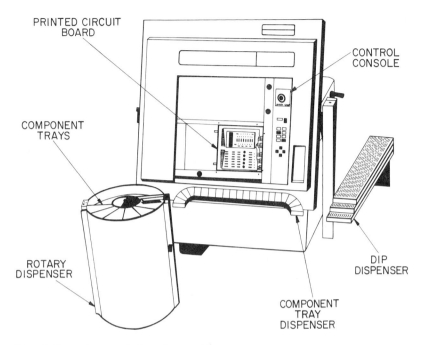

PRINTED CIRCUIT
BOARD

CONTROL
CONSOLE

COMPONENT
TRAYS

ROTARY
DISPENSER

DIP
DISPENSER

COMPONENT
TRAY
DISPENSER

Fig. 9.8 Semiautomatic insertion machine.

after, *wave soldering*. The two stations are sometimes needed because some components cannot withstand the high temperatures of wave soldering. Both are manual insertion stations, which may involve the use of special hand tools to facilitate the handling and insertion of certain components. Typically, manual assembly accounts for a high proportion of total assembly time, even when a relatively small number of manually assembled components are involved.

At the first manual assembly station, large nonstandard components are inserted. They are inserted manually because the part trays used in semiautomatic machines cannot accommodate many of these parts on account of their large size. If a particular manufacturer does not use a semiautomatic machine, all the components that would have been inserted using this machine will be inserted manually. Also, components assembled mechanically (i.e., secured with screws or bolts) that are to be wave-soldered are assembled here.

After the first manual assembly station is a block indicating retrofit (Fig. 9.1). Assembly here is also manual but involves only engineering change order (*ECO*) wires or jumper wires. These wires are cut to the required length from spools, and the ends are stripped and *tinned*. One end of the wire is soldered to a component lead on top of the board or is inserted through the board; the wire is routed around the board and *cemented* down at various points; finally, the other end is soldered to a component lead or inserted into the board. These ECOs are often needed to satisfy certain customer options and are sometimes needed because it is difficult to fit the entire etched circuit on a board without having to cross paths at some point. An alternative is to use a multilayer circuit board (one with two or more circuits printed on it with insulating material separating the circuits).

An assembly station for robotic insertion is included in the assembly process (Fig. 9.1) and occurs after retrofit in the flow diagram. At this station, robotic insertion of nonstandard components using a one-arm robot may be performed. Robot insertion is used mainly to reduce the amount of manual labor involved in PCB assembly.

Solder-clean, the next station in the assembly process (Fig. 9.1), refers to wave-soldering of the component leads on the underside or solderside of the board and the removal of excess flux applied to the solderside of the board prior to soldering. A conveyor carries the PCB assembly over a rounded crest or wave of solder so that the board impinges against the wave in passing. The purpose of applying a flux, usually an organic rosin type, is to remove from the board oxides that inhibit solderability. Cleaning is subsequently done to remove excess flux, which could cause corrosion and/or contamination.

The time taken for wave-soldering and cleaning one board or, in the case of small boards, one panel or fixture, depends on the conveyor speed, with conveyor speed adjustment up to 20 fpm. Typically, conveyor speeds are near 10 fpm, which yields a time of approximately 2 min for a board, panel, or fixture to pass through the wave-solder and cleaning station. However, conveyor speeds are selected to yield specific solder contact times, which are usually about 3 sec.

Immediately following solder and clean is *touch-up*. Touch-up is cleaning the solderside of the board to remove excess solder. This is necessary because of the tendency of the solder to icicle and bridge, which can cause short circuits in the electrical layout. Icicling and bridging are more prevalent on closely packed boards.

At the second manual assembly station (often referred to as "final assembly"), all the remaining components are inserted (Fig. 9.1). These include components that cannot be wave-soldered because of their sensitivity to heat. Also, components that are secured mechanically are installed here. These include handles, some large electrolytic capacitors, connectors, and power *transistors*, which are secured with screws or bolts and nuts. Lastly, some components, such as diodes and resistors, usually with axial leads, are soldered on the top of the board to the leads of other components. These operations, if necessary, are a type of engineering change order, and there is a need to reduce the number of components inserted at this station because of the greater time involved with hand-soldering compared to wave-soldering.

9.4 SMD TECHNOLOGY

The above discussion has dealt only with PCBs having through-hole components where the component leads pass through the board. However, the surface-mount devices (*SMDs*) are increasingly employed. These components have pads or leads soldered to corresponding areas on the surface of the board. Assembly of an SMD involves positioning the component on the board, which has previously had the solder paste applied. When all the SMDs have been positioned, the board is *reflow-soldered*. This involves heating the solder paste until it flows into a uniform solder layer, which permanently affixes the SMD pads or leads to the board. Passive SMD components such as resistors and capacitors can also be added to the underside of the board using adhesive. Soldering them takes place during wave solder.

Surface-mount devices include simple resistors in the form of a small rectangular prism and a variety of larger components, such as flat packs (Fig. 9.2). These devices can be mounted on either side of the board and are usually interspersed with through-hole components.

Also, there are many variations on the PCB assembly sequence described here. In particular, the use of robots is increasing and, for high-volume production, these can be arranged in assembly-line fashion, as illustrated in Fig. 9.9. With such an arrangement, the components are usually presented in standardized *pallets*, roughly oriented. The robot must then provide the final orientation prior to insertion and, for this purpose, vision systems might be employed. Alternatively, some components or mechanical parts are delivered to a station using standard feeding and orienting techniques. The robot then acquires the preoriented component or part at the station.

It should be realized that PCB manufacture is a rapidly developing process. New PCB designs, new component packaging, and new assembly techniques are continually being introduced. Manufacturers constantly strive to achieve higher component densities, which makes assembly increasingly difficult.

9.5 ESTIMATION OF PCB ASSEMBLY COSTS

The materials necessary for an estimation of PCB assembly costs are presented at the end of this section. Data bases giving the times for manual insertion are included, and a worksheet is provided to assist in tabulating the results. Components and operations are entered on the worksheet in assembly order; one line for each basic type of component or operation.

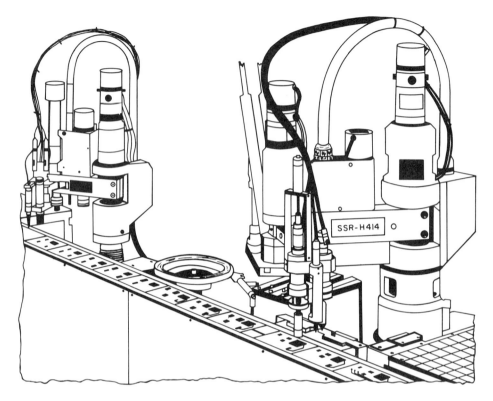

Fig. 9.9 Robot assembly of printed circuit boards.

The time for manual insertion, obtained from the data base, is entered on the worksheet and then multiplied by the operator rate to give the insertion cost. After adding per-component allowances for *rework* costs, the total operation cost is entered.

For automatic or robot insertion, the cost is obtained directly from the data base and then adjusted for programming, setup, and rework.

For mechanical parts, the manual assembly times and costs can be obtained using the *Product Design for Assembly Handbook* [2]. When all the operations have been entered, the total cost is obtained by summing the figures in the cost column.

Figure 9.10 shows, by way of example, a completed worksheet for a PCB assembly taken from a microcomputer. Such an assembly is commonly referred to as a "logic board;" this particular board contains 69 DIPs, 1 DIP socket, 16 axial components, and 32 radial components. In addition, two parts are attached mechanically, requiring 11 screws, nuts, and washers. It is

Name of PCB COLOR CARD	Operator rate W_a=0.6 cents/s		Batch size, B_s=1,000	No. of setups N_{set}=_10_		No. of boards/ panel N_b=_2
Name of part, sub-assembly operation or soldered component	No. of pins or leads N_1	No. of parts ops. comps. R_p	Total manual op.time (sec) T_a	Total op.cost (cents) C_a	Figs. for min. parts N_m	Description
Circuit board	-	1	4.0	2.4	1	place in fixture
DIP	20	11	-	19.9	-	auto. insert
DIP	14	56	-	82.9	-	auto. insert
axial	2	16	-	24.3	-	auto. insert
radial	2	32	-	47.0	-	auto. insert
DIP socket	24	1	18.0	13.0	-	man. insert
DIP	40	1	26.0	19.2	-	man. insert
screw	-	2	15.6	9.4	0	add
coax. connector	3	1	8.0	5.1	-	man. insert
display connector	9	1	8.0	5.6	-	man. insert
star washer	-	2	17.4	10.4	0	add & hold
nut	-	2	8.8	5.3	0	add & screw
end plate	-	1	4.5	2.7	1	add
lock washer	-	2	9.4	5.6	0	add
hex. nut	-	2	19.8	11.9	0	add & screw
screw	-	1	9.8	5.9	0	add & screw
wave solder	-	1	-	75.0	-	wave solder
DIP	24	1	4.0	2.4	1	add & snap fit
			Total	348.0		

Fig. 9.10 Completed worksheet for sample PCB assembly.

assumed that the DIPs, radials, and axials are autoinserted and that the remaining components or parts are manually inserted or assembled except for one DIP, which is assembled into the corresponding DIP socket after wave solder. The completed worksheet for this example board gives a total estimated assembly cost of $3.48. Consideration of avoidable costs indicates that elimination of the 11 fasteners would save 48.0 cents, a surprisingly high figure for a board with only one nonelectrical component, namely, the end plate.

The operations costs for the automatic insertion processes include the cost of rework, which amounts to a total of 46.1 cents—clearly a significant item.

In estimating the cost of wave solder, it was assumed that two boards would be processed together. However, the resulting use of 75 cents is a significant item and, in practice, should be examined closely for accuracy.

Finally, it was assumed in this analysis that the manufacturer would have an automatic machine available for the insertion of the 32 radial components. It is interesting to note that if these components were to be inserted manually, an additional expense of $2.19 would be incurred, increasing the total cost of assembly by 63%.

9.6 WORKSHEET AND DATA BASE FOR PCB ASSEMBLY COST ANALYSIS

Instructions:

1. For soldered components, assemblies, or operations, use the data base provided here. For all other parts, components, or operations, use the manual assembly data from the *Product Design for Assembly Handbook*.[2]
2. Record the data on the worksheet in the following order of assembly:
   a. Load PCB into fixture.
   b. Insert all wave-soldered components.
      (i) autoinserted
      (ii) robot-inserted
      (iii) manually inserted
   c. Wave-solder.
   d. Insert and solder all manually-soldered components:
      (i) autoinserted
      (ii) robot-inserted
      (iii) manually inserted
   e. Insert and secure all remaining nonsoldered components.

If parts and mechanical fasteners are associated with any soldered components, list them with the appropriate component.

9.7 PCB ASSEMBLY—EQUATIONS FOR TOTAL OPERATION COST

Manual:

$$C_{\text{op}} = t_a W_a + R_p C_{\text{rw}} M_f \text{ cents}$$

PCB Assembly Worksheet

Name of PCB _____	Operator rate W_a=_._ cents/s	Batch size, B_s=_____	No.ofsetups N_{set}=___	No. of boards/ panel N_b=___

Name of part, sub-assembly operation or soldered component	No. of pins or leads N_1	No. of parts ops. comps. R_p	Total manual op.time (sec) t_a	Total op.cost (cents) C_{op}	Figs. for min. parts N_{min}	Description
			Total			

NOTE: t_a is inappropriate for auto or robot insertion unless hand solder is carried out; N_{min} is inappropriate for soldered electronic components.

PCB Assembly Database - Manual Operations

Insertion of Components		
Components Types	Operations	Time(s)
Axials (VCDs)	bend leads, insert, cut and clinch leads	19.0
Radials or can-type ICs	insert component cut and clinch leads	basic time	10.0
		additional time per lead	1.8
SIP/SIP sockets or connectors	insert component	<= 80 leads	8.0
		> 80 leads	10.0
Posts, DIP/DIP sockets, pin-grid arrays or odd-form components	insert component	basic time	6.0
		additional time per lead or post	0.5
SMDs	add and solder using special fixture or tool	10.0

Other Manual Operations		
Part	Operation	Time(s)
Sleeve	cut one sleeve and add to a lead	15.0
Jumper wire (ECO)	cut, strip and tin ends, insert and solder	60.0
Heat sink	add to transistor	25.0
Lead or post	hand-solder one lead or post	6.0

NOTE: These data have largely been adapted from Ref. 4.

Autoinsertion machine:

$$C_{op} = R_p(C_{ai} + C_{ap}/B_s + C_{rw}A_f) + N_{set}C_{as}/B_s \text{ cents}$$

Robot insertion machine:

$$C_{op} = R_p(C_{ri} + C_{rp}/B_s + C_{rw}R_f) + N_{set}C_{rs}/B_s \text{ cents}$$

PCB Assembly Database - Automatic Operations

Insertion Costs per Component (cents)		
Component Type	Auto, C_{ai}	Robot, C_{ri}
Axial (VCD)	1.2*	5.0
Radial	1.2*	5.0
SIP/SIP socket	0.8	5.0
DIP/DIP socket	0.8	5.0
Connector	1.0	5.0
Small SMD (2 connections)	0.2	5.0
Large SMD (>2 connections)	1.0	5.0

Associated Costs (dollars)	
Wave or reflow-solder**	$1.50

*Includes cost of sequencing.

**If several boards are contained in one panel or secured in a single fixture, divide this figure by N_b, the number of boards per panel or fixture, in order to obtain the cost of wave or reflow solder per board. This figure also includes the time to place the boards or panel in the soldering fixture.

where A_f = average number of faults requiring rework for each auto insertion (0.002)

B_s = total batch size

C_{ai} = cost of autoinsertion (obtained from data base)

C_{ap} = programming cost per component for autoinsertion machine (150 cents)

C_{as} = setup cost per component type for autoinsertion machine (150 cents)

C_c = cost of replacement component when rework is needed

C_{ri} = cost of robot insertion (obtained from data base)

C_{rp} = programming cost per component for robot system (150 cents)

C_{rs} = setup cost per component type for robot system (150 cents)

C_{rw} = $T_{rf}N_lW_a + C_c$ and is the average rework cost per faulty component (cents)

M_f = average number of faults requiring rework for each manual insertion (0.005)

N_ℓ = number of leads or posts on one component

N_{set} = estimated number of setups per batch

R_f = average number of faults requiring rework for each robot insertion (0.002)

R_p = number of components

t_a = total time for all manual operations (calculated fom data base
figures)

T_{rf} = average estimated time to rework one component fault per lead or
post (30 sec)

W_a = rate for manual operations, cents/sec

9.8 GLOSSARY OF TERMS

The following is a brief description of some of the terminology used in this
chapter. The physical configurations of some electrical components, the
manual assembly operations, and some terms pertaining to automatic insertion
equipment are described.

Axial-Leaded Component

Electrical component of cylindrical shape with two leads exiting from opposite
ends of the component in line with its axis (Fig. 9.2). The component is sometimes called a VCD (variable center distance). The most common axial-leaded
components are resistors, capacitators, and diodes.

Can-Type IC

A cylindrical integrated circuit, packaged so that the leads form a circular pattern (Fig. 9.2). This multileaded radial component can have 3–12 leads.

Cement

Because of the requirements for ruggedness placed on some PCBs, components can be cemented to the board to reduce the effects of vibrations. This
sometimes requires that the component leads be bent prior to cementing.

Center Distance

Distance between leads when formed for insertion. This term applies to two-
leaded components and DIPs; also termed *lead span*.

Channel

A plastic container, in the form of a long tube, in which a number of DIPs are
placed in single file and oriented for dispensing to an insertion machine; also
called *sticks*.

Clinch

The bending of a component lead end after insertion through the PCB. This
temporarily secures the component prior to soldering. In a full clinch, the lead
is bent to contact the terminal area. In a modified clinch, the lead is partially
bent to a predetermined intermediate angle.

Component
Any electrical device to be attached to the PCB.

DIP
Dual in-line package (Fig. 9.2). A rectangular integrated circuit, packaged so as to terminate in two straight rows of pins or leads.

ECO
Engineering change order. A component or insulated jumper wire, installed manually, that is needed when the electrical circuit cannot be etched onto the board without crossing paths at some point. Often the leads are not inserted into the board but are manually soldered to the leads of components already assembled to the board. ECO wires are also referred to as *jumper wires*. (See retrofit).

Fault
Any error that causes the assembled PCB to fail testing procedures and that requires rework.

Hybrid
A PCB populated by both surface-mount and through-hole components; also referred to as *mixed-mounting* technology.

Insertion
Process whereby the component is grasped, prepared if necessary, placed on the board, and temporarily secured if necessary.

Kitting
Preparing a package of parts, usually with instructions for assembly, to facilitate manual assembly.

Magazine
A unit containing a group of dispensing channels, usually about 15, used for an automatic DIP insertion machine.

Module
A unit containing a group of dispensing spools, usually 20 or less, used with an automatic axial-lead component sequencer.

Nonstandard component
Any component that cannot be inserted by dedicated automatic machinery because of its physical characterisitcs, i.e., size, shape, lead span, etc., also called *odd form* components.

Pallet
A tray on which components are arranged in a known position and orientation.

PCB

Printed-circuit-board. An insulating board onto which an electrical circuit has been printed in the form of conductive paths; contains drilled holes into which the leads of components are inserted; also known as a *printed wiring board (PWB)*.

Preform

Forming the leads of a component to the correct dimensions prior to insertion. Axial-leaded components must usually have their leads bent at right angles for insertion, and DIPs sometimes require lead or pin straightening. Radial-leaded components may have their leads notched or a stand-off or spacer installed, which maintains the required clearance between the component and the board. Can-type ICs and transistors often need a type of lead forming called *form a*, which refers to the profile of the leads after forming.

Radial-Lead Component

Electrical components with leads at right angles to the body (Fig. 9.2). Examples are disk capacitators, "kidney" or "jellybean" capacitators, cermet resistors, etc.

Reflow Solder

Process by which surface-mount devices become secured to the PCB.

Retrofit

A type of ECO that involves only the assembly of wires (jumper wires) to the PCB; can refer to an assembly station in the PCB assembly process where only wires are assembled.

Rework

Repair a fault. This usually means severing the leads of the component and removing it, removing the individual leads from the PCB holes, cleaning the holes, inserting a new component, and soldering its leads. The operations are performed manually and are time-consuming and expensive.

Sequencer

A machine that arranges components on a tape spool in the correct order for insertion.

SIP

Single in-line package. An integrated circuit, usually a resistor network or a connector, packaged so as to terminate in one straight row of pins or leads (Fig. 9.2).

Sleeve

An insulating plastic tube slipped manually onto the lead of a component prior to insertion to guard against electrical short circuits (Fig. 9.2).

SMD

Surface-mount device. A component (often leadless) that is secured to the surface of the board.

Spacer

This can be a small plastic ring (Fig. 9.2) used to keep a minimum clearance between the component and the board. It is usually cemented to the board before the component is inserted. Some components use temporary spacers that are removed after the component is secured. Some spacers are provided with holes corresponding to each lead (Fig. 9.2).

Spool

The package for holding taped axial-leaded components.

Standard Component

Any component that can be inserted by an automatic insertion machine.

Stick

A plastic container in which a number of DIPs are aligned in single file and are oriented for dispensing to an automatic insertion machine.

Surface Mount

See SMD.

Tin

Provide a layer of solder on the surface of leads prior to insertion.

Touch-up

Cleaning the underside or solderside of a PCB after wave soldering to remove any excess solder, which can cause short circuits.

Transistor

A small component whose body has a cylindrical envelope except for one flat face, with three leads at right angles to the body (Fig. 9.2).

VCD

Variable center distance. The capability of an axial-lead component insertion head to vary the distance between leads when forming and inserting an axial-lead component (Fig. 9.2). The term is also used to refer to an axial-lead component. The terms *adjustable span* and *variable span* can also be used.

Wave-Solder

To solder automatically all the leads on an assembled PCB by conveying it, at a slight incline, over a wave of solder.

REFERENCES

1. John, J., and Boothroyd, G., "Economics of Printed Circuit Board Assembly," Report 6, Economic Application of Assembly Robots Project, Univ. of Massachusetts, Amherst, April 1985.
2. Boothroyd, G., and Dewhurst, P., *Product Design for Assembly Handbook*, Boothroyd Dewhurst, Inc., Wakefield, RI, 1987.
3. Russell, G.A., Boothroyd, G., and Dewhurst, P., "Printed Circuit Board Design for Assembly," Report 5, Product Design for Manufacture Program, Department of Industrial and Manufacturing Engineering, Univ. of Rhode Island, Kingston, R.I., May 1986.
4. Boothroyd, G., and Shinohara, T., "Component Insertion Times for Electronics Assembly," *International Journal of Advanced Manufacturing Technology*, 1 (5), 1986, p. 3.

10

Feasibility Study for Assembly Automation

The subject of this chapter brings together the results of work in all the various aspects of assembly. After a careful analysis of the product design, the designer of an assembly machine must produce a proposal that combines many requirements. Some of these requirements, such as reliability and durability, are similar to those for any machine tool. However, certain requirements are applicable only to the assembly machine and are mainly a result of the variations in the quality of the component parts to be assembled. It can reasonably be assumed that an assembly machine can be designed that, if it is fed only with carefully inspected parts, will repeatedly perform the necessary assembly operations satisfactorily. Sometimes, unfortunately, the real problems of automatic assembly appear only when the machine is installed in the factory and the feeders are loaded with production parts containing the usual proportion of defectives. The possible effects of feeding a defective part into an assembly machine are:

1. The mechanical workhead may be seriously damaged, resulting in several hours or even days of downtime.

2. The defective parts may jam in the feeder or workhead and result in machine or workhead downtime while the fault is cleared.

3. The part may pass through the feeder and workhead and spoil the assembly, thus effectively causing downtime equal to one machine cycle and producing an assembly that must be repaired.

Often, large samples of the parts to be assembled are available when the assembly machine is designed and, clearly, the designer should take into account, at the design stage, the quality levels of the parts and the possible difficulties resulting from them. Even the choice of basic transfer system can significantly affect the degree of difficulty caused by defective parts and, therefore, most of the important decisions will be made by the machine designer before detailed design is considered. Basically, the object of design should be to obtain minimum downtime on the machine resulting from a defective part being placed in the corresponding feeding device. This was illustrated in Chapter 6, which dealt with the economics of multistation indexing assembly machines. It was shown that any reductions in the time taken to clear a fault caused by a defective part will reduce the assembly costs by reducing the total machine downtime. The first section of this chapter discusses the various design factors that can help to ensure minimum downtime due to defective parts, and the second section describes a feasibility study in detail.

10.1 MACHINE DESIGN FACTORS TO REDUCE MACHINE DOWNTIME DUE TO DEFECTIVE PARTS

The first objective in designing feeders and mechanisms for use in automatic assembly is to ensure that the presence of a defective part will not result in damage to the machine. This possibility does not generally exist where the part is moving under the action of its own weight (that is, sliding down a chute) or being transported on a vibrating conveyor. However, if the part is being moved or placed in a positive way, it is necessary to arrange that the desired motion is provided by an elastic system. In this case, if a defective part becomes jammed, motion can be taken up in the spring members. For example, if a plunger is to position a part in an assembly, it would be inadvisable to drive the plunger directly by a cam. It would be better to provide a spring to give the necessary force to drive the plunger and use the cam to withdraw the plunger. With this arrangement, a jammed part could not damage the mechanism.

The next objective in design should be to ensure that a jammed part can be removed quickly from the machine. This can be facilitated by several means, some of which are as follows:

1. All feeders, chutes, and mechanisms should be readily accessible. External covers and shields should be avoided wherever possible.

2. Enclosed feed tracks, feeders, and mechanisms should not be employed. Clearly, one of the least expensive forms of feed track is a tube down which the parts can slide freely to the workhead. However, a jam occurring in a closed tube is difficult to clear. Although probably more expensive to provide, open rails are preferable in this case so that the fault can be detected and cleared quickly.

3. An immediate indication of the location of a fault is desirable. This may be achieved by arranging that a warning light is switched on and a buzzer sounded when any operation fails. If the warning light is positioned at the particular workhead, the technician will be able to locate the fault quickly.

It is necessary for the machine designer to decide whether to arrange that the machine is stopped in the event of fault or whether to arrange that the spoiled assembly continue through the machine. In Chapter 6, we saw that, in typical circumstances, it is preferable to keep the machine running. However, it would be clearly undesirable to attempt further operations on the spoiled assembly. For this purpose, the memory pin system can be employed, where each work carrier is fitted with a pin which, in the event of a failure, is displaced by a lever fitted to the workhead. Each workhead is also provided with a feeler that senses the position of the pin before carrying out the operation. If the pin is displaced, the operation is not carried out. A microprocessor can accomplish this same effect with appropriate software and electronic control.

The difficulty with the memory pin system is that it is not possible for the workhead to detect immediately whether a particular fault will be repetitive. One possibility is to arrange that, initially, any type of fault will displace the memory pin but that when two or three faults have occurred in succession, the machine is automatically stopped. Alternatively, it may be left to a technician to observe when a succession of faults occurs and then to stop the machine.

The discussion above dealt with methods of reducing machine downtime caused by defective parts. Ideally, of course, the defective parts are detected and rejected in the feeding devices. Although it is generally not possible to perform complete inspection during the feeding of parts, it is sometimes possible to eliminate a considerable proportion of the defective parts. Figure 10.1 shows an example given by Ward [1] in which unsawn bifurcated rivets are detected and rejected in a bowl feeder. In this case, the device was incorporated to prevent unsawn rivets from being fed to the riveting head, where they would damage the mechanism during the operation. Sometimes, small pieces of swarf and other foreign bodies find their way into a bowl feeder, and a simple way of rejecting these is illustrated in Fig. 10.2. Quite sophisticated inspection arrangements can be built into a bowl feeder, and it has been found economical in some circumstances to develop such a device to inspect the parts before they are placed in the assembly machine feeders. In this case, the downtime occurs on the inspection device instead of on the assembly machine.

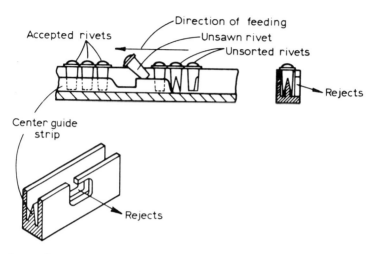

Fig. 10.1 System for inspecting bifurcated rivets (from Ref. 1).

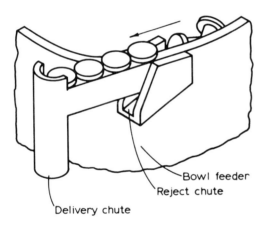

Fig. 10.2 System for rejecting foreign matter in vibratory-bowl feeder (from Ref. 1).

10.2 FEASIBILITY STUDY

The decision to build or purchase an automatic assembly system is generally based on the results of a feasibility study. The object of this study is to predict the performance and economics of the proposed system. In automatic assembly, these predictions are likely to be subject to greater errors than with most other types of production equipment, mainly because the system is probably one of a kind and its performance depends heavily on the qualities of the parts to be assembled. Also, similar systems will not generally be available for study. Nevertheless, a feasibility study must be made, and all the knowledge and experience acquired in the past from automatic assembly projects must be applied to the problem in order to give predictions that are as accurate as possible.

Certain information is clearly required before a study can be made. For example, maximum and minimum production rates during the probable life of the machine must be known. The range of variations in these figures is important because a single assembly machine is quite inflexible. The personnel required on the machine must all be present when the machine is working or, if the machine is stopped because of a falloff in demand for the product, they must be employed elsewhere. Thus, automatic assembly machines are generally suitable only when the volume of production is known to be steady. Further, they can usually be applied profitably only when the volume is high. Apart from this high-volume requirement, the labor costs of the existing assembly process must also be high if automatic assembly is to be successful.

Clearly, much more information will be required and many other factors will combine to determine the final answer. Some of these are discussed in greater detail below.

10.2.1 Precedence Diagrams

It is always useful when studying the assembly of a product to draw a diagram that shows clearly and simply the various ways in which the assembly operations may be carried out. In most assemblies, there are alternatives in the order in which some of the parts may be assembled. There are also likely to be some parts for which no flexibility in order is allowed. For example, in the three-pin power plug shown in Fig. 10.3, the pins may be placed in position in any order, but the fuse can be inserted only after the fuse clip and live pin are in position. Further, the cover can be placed in position and secured only after all the remaining parts have been assembled into the base. The precedence diagram is designed to show all these possibilities, and its use has been described in detail by Prenting and Battaglin [2]. A precedence diagram for the assembly of the power plug, assuming that no subassemblies are involved, is shown in Fig. 10.4, where it can be seen that each individual operation has been assigned a number and is represented by an appropriate circle with the number inscribed. The circles are connected by arrows showing the precedence relations.

In the precedence diagram, all the operations that can be carried out first are placed in column I. Usually, only one operation appears in this column:

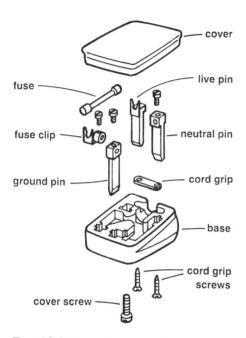

Fig. 10.3 Three-pin power plug.

the placing of the base part on the work carrier. Operations that can be performed only when at least one of the operations in column I has been performed are placed in column II. Lines are then drawn from each operation in column II to the preceding operations in column I. In the example in Fig. 10.4, none of the column II operations can be performed until the base of the power plug has been placed on the work carrier and, therefore, lines are drawn connecting operations 1a, 2, 3, 4, and 5 to operation 1. Third-stage operations are then placed in column III, with appropriate connecting lines, and so on, until the diagram is complete. Thus, following all the lines from a given operation to the left indicates all the operations that must be completed before the operation under consideration can be performed.

In the assembly of the power plug, there are 15 operations, and it will probably be impracticable to carry out all these on a single machine. For example, it would be difficult to assemble the cord grip and the two cord grip screws while the plug base is held in a work carrier because these parts enter the base from different directions. It is probably better, therefore, to treat these parts as a subassembly, and this is indicated in Fig. 10.4 by the dashed line enclosing the necessary operations. In a similar way, the neutral and earth pins and the fuse clip, together with their respective screws, can also be treated as

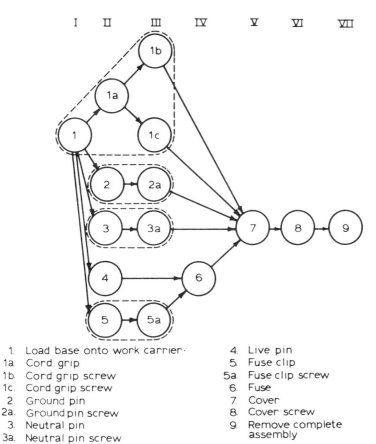

| I | II | III | IV | V | VI | VII |

1. Load base onto work carrier·
1a. Cord grip
1b. Cord grip screw
1c. Cord grip screw
2. Ground pin
2a. Ground pin screw
3. Neutral pin
3a. Neutral pin screw

4. Live pin
5. Fuse clip
5a. Fuse clip screw
6. Fuse
7. Cover
8. Cover screw
9. Remove complete assembly

Fig. 10.4 Precedence diagram for complete assembly of power plug.

subassemblies. These groups of operations are all indicated by the dashed lines in Fig. 10.4.

One of the objectives in designing an automatic assembly machine should be to include as few operations as possible on the line in order to keep machine downtime to a minimum. It is desirable, therefore, to break the product down into the smallest number of subassemblies and carry out individual studies of the subassemblies. If these subassemblies can be performed mechanically, separate machines may be used. These machines can then be arranged to feed the main assembly machine at the appropriate points.

Figure 10.5 shows the precedence diagram for the subassemblies of the power plug. It can be seen that no flexibility exists in the ordering of opera-

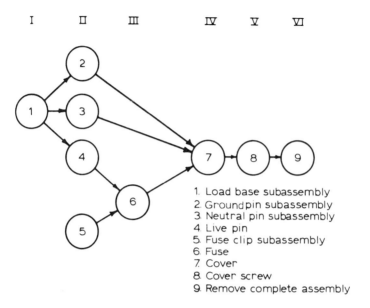

I   II   III   IV   V   VI

1. Load base subassembly
2. Groundpin subassembly
3. Neutral pin subassembly
4. Live pin
5. Fuse clip subassembly
6. Fuse
7. Cover
8. Cover screw
9. Remove complete assembly

Fig. 10.5 Precedence diagram for assembly of power plug subassemblies.

tions 1, 7, 8, and 9. Operations 2, 3, 4, 5, and 6, however, can be carried out in any order between operations 1 and 7 except that 6 cannot be performed until both 4 and 5 are completed. Considering the group of operations 4–6 first, there are two ways in which these can be performed: either 4, 5, 6 or 5, 4, 6. Operation 3 could be performed at any stage in this order, giving $4 \times 2 = 8$ possibilities. Thus, the precedence diagram shown in Fig. 10.5 represents 40 possible orderings of the various assembly operations and will be useful when we consider the layout of the proposed assembly machine.

10.2.2 Manual Assembly of Plug

Before embarking on a detailed study for automatic assembly, it is necessary to analyze the product for manual assembly. This provides not only a bench-mark for economic justification but also information on manual handling and assembly for any parts that do not lend themselves to automatic handling and assembly.

Table 10.1 presents the results of a manual assembly analysis carried out using the procedures described in Chapter 7. The total manual assembly time is 43 sec, with a corresponding cost of 36 cents for an assembly worker rate of $30/hr. With an efficiency of 80% for the assembly plant, the manual assembly cost would be 45 cents.

Table 10.1 Analysis of Power Plug for Manual Assembly

	Handling code	Handling time, sec	Insertion code	Insertion time, sec	Total time, sec	Cost, cents	Minimum parts
1 Base sub	30	1.95	00	1.5	3.5	2.9	1
2 Fuse clip sub	35	2.73	00	1.5	4.2	3.5	1
3 Live pin	20	1.80	00	1.5	3.3	2.8	1
4 Fuse	00	1.13	31	5.0	6.1	5.1	1
5 Ground pin sub	20	1.80	00	1.5	3.3	2.8	1
6 Neutral pin sub	20	1.80	00	1.5	3.3	2.8	1
7 Cover	30	1.95	06	5.5	7.4	6.2	1
8 Reorientation	—	—	98	4.5	4.5	3.8	—
9 Cover screw	10	1.50	38	6.0	7.5	6.3	0
Totals					43.2	36.2	7

Design efficiency = $7 \times 2.88/43.1 = 47\%$.

The design efficiency for the plug is very high at 47%, mainly because all the items except the cover screw must, of necessity, be separate from one another.

10.2.3 Quality Levels of Parts

If assembly of a completely new product is to be contemplated, the estimation of the quality levels of the parts may be extremely difficult, if not impossible. However, a large proportion of assembly machine feasibility studies are concerned with existing products and, in these cases, experiments can be performed to determine the quality levels of the various parts. It should be remembered in such a study that defective parts do not generally create great difficulties when assembly is performed by hand. Often, the assembly worker can quickly detect and reject a defective part and, in many cases, when the "defective part" is simply a nonpart, such as a piece of swarf or bar end, the assembly worker does not even attempt to grasp it but simply leaves it in the parts container to be discarded later. This means that a study of quality level must be conducted at the existing assembly stations, where the numbers of discarded parts and foreign bodies can be recorded. A further danger is that engineers responsible for assembly processes often assume that 100% visual inspection results in 100% acceptable parts. The assumption that an inspection worker inspecting every part that is to be assembled will detect every defective part is clearly not valid.

The best procedure for estimating quality levels is for the investigator to observe the assembly work and note every defective part or foreign body that is discarded. Obviously, it is inadvisable to assume that the quality levels recorded cannot be improved upon, but it is necessary to estimate the cost of these improvements and to allow for this extra cost in the feasibility study.

Having noted the number of defective parts in a given batch, the investigator can then divide these into two categories: (1) those parts that cannot be assembled, for example, screws with no thread or slot; and (2) those parts that can be assembled but are normally rejected by the operator, for example, discolored or chipped parts.

The number of parts falling within the first category allows estimates to be made of the assembly machine downtime, and the number of those falling within the second category allows estimates to be made of the number of unacceptable or defective assemblies produced by the machine.

Figures for the power plug shown in Fig. 10.3 are presented in Table 10.2. It is important to remember that no assessment can be made of the most suitable type of assembly machine or of the number of operations that can economically be performed mechanically until the individual quality levels of the various parts have been investigated.

Table 10.2 Quality Levels of Power Plug Parts

Parts	Fault	Number of faults in assembling 10,000 plugs	Percentage faults
Base	Chipped	10	0.10
subassembly	Earth pin will not assemble	170	1.70
	Live pin will not assemble	20	0.20
	Neutral pin will not assemble	30	0.30
Earth pin subassembly	No screw	41	0.41
Neutral pin subassembly	No screw	59	0.59
Live pin	Fuse will not assemble	123	1.23
	Fuse assembles unsatisfactorily	21	0.21
Fuse clip	Fuse will not assemble	115	1.15
subassembly	Fuse assembles unsatisfactorily	17	0.17
Fuse	Damaged	18	0.18
Cover	Chipped	10	0.10
	Cover screw hole blocked	200	2.00
Cover screw	No thread or slot	20	0.20

10.2.4 Parts Feeding and Assembly

An estimate must now be made of the degree of difficulty with which the individual parts can be automatically fed and assembled. It should be noted here that, for each operation, four possibilities exist:

1. Automatic feeding and assembly
2. Manual feeding and automatic assembly
3. Automatic feeding and manual assembly
4. Manual feeding and assembly

For this feasibility study, it was assumed that an annual production volume of 1.5 million plugs is planned with two-shift working. For a product life of 36 months, this represents a total life volume of 4.5 million assemblies for an efficiency of 80%. An assembly rate of 7.8/min would be required while the assembly system is in operation, giving an average production time or cycle time of 7.7 sec. This figure includes any downtime due to faulty parts.

In considering the feeding of parts, all but the simplest shapes will probably require vibratory-bowl feeders, and simple experiments can normally be performed to test various ideas for orienting and feeding. Estimates can then be made of the various feed rates possible. For a given bowl feeder, the maximum feed rate obtainable is proportional to the reciprocal of the length of the part, assuming that all parts arrive at the bowl outlet end to end. Thus, with large parts that have many possible orientations, only one of which will be required, the feed rate of oriented parts can be very low.

For example, the base subassembly of the power plug shown in Fig. 10.3 is 49 mm long and can be fed along a bowl feeder track in any of eight possible orientations. Thus, suitable devices can be fitted to the track of the bowl to reject seven of the eight orientations. The maximum feed rate of oriented parts will be given by Eq. (3.49). Thus,

$$F = \frac{vE}{\ell} \qquad (10.1)$$

where v is the mean conveying velocity (25 mm/sec), ℓ the length of the part, and E the modified efficiency of the orienting system [see Eq. (3.50)]. From the *Product Design for Assembly Handbook* [3], the automatic handling code for the base subassembly is obtained as follows:

1. The overall dimensions of this nonrotational subassembly are $A = 49$, $B = 47$, $C = 12$ mm. Thus, $A/B < 3$, and $A/C > 4$. From Fig. 8.3, therefore, the first digit of the geometrical code is 6.

2. The part has no symmetry about any axis, and the principal features that can be used for orienting purposes are the two screws that appear as an asymmetric projection or step when the subassembly is viewed in the Y direction. Thus, from Fig. 8.5, the second and third digits are 4 and 1, respectively.

3. In a vibratory-bowl feeder, parts circulate and tumble from the tracks into the bottom of the bowl repeatedly. This part is manufactured from hard plastic, and care must be taken not to damage its surfaces. For a delicate part such as this, Fig. 8.8 gives fourth and fifth digits of 2 and 0, respectively.

4. The five-digit code is, therefore, 64120 giving, from Fig. 8.5, an orienting efficiency of 0.15 and a relative feeder cost of 1. With an additional feeder cost of 1 from Fig. 8.8, we get a total relative feeder cost of 2.

Using Eq. (10.1), the estimated maximum feed rate from one feeder would be

$$F_m = 49 \times \frac{0.15}{25} = 4.6 \text{ parts/min}$$

To meet the required feed rate of 7.8 parts/min, two feeders would be required or, alternatively, a sum equivalent to the cost of 1.7 feeders could be spent on the development of a special feeder to meet the required delivery rate. In the present case, if the basic feeder cost is $7k, the total sum would be $23.8k.

Bearing these points in mind, the designer decides whether automatic feeding of the particular part is feasible. In the example of the power plug, it is possible that both the base subassembly and the cover could not be fed at the required rate from one feeder and that the fuse clip assembly, because of its complicated shape, should be handled manually. The remaining parts and subassemblies could probably all be fed and assembled automatically with bowl feeders and placing mechanisms, with the exception of the main holding screw. This could be fed and screwed from below with a proprietary automatic screwdriver.

10.2.5 Special-Purpose Machine Layout and Performance

Indexing Machine

If it is assumed that the base, top, and fuse clip are to be assembled manually on an in-line indexing machine, at least two assembly workers will be required. The first, positioned at the beginning of the line, could place the base subassembly on the work carrier and place the fuse clip assembly in the base (operations 1 and 5 of Fig. 10.5, respectively). The second assembly worker could assemble the cover and remove the complete plug assembly from the end of the line (operations 7 and 9 of Fig. 10.5, respectively).

It is generally necessary on an assembly machine to include some inspection stations. In the present example, it is clear that after the plug cover has been assembled, there will be no simple means of inspecting for the presence of the fuse clip, the fuse, and the three small screws in the neutral and earth pins and the fuse clip. Thus, it will be necessary to include an inspection head on the machine immediately before operation 7 (the assembly of the cover), which will check for the presence of all these parts.

In the present example, it is also necessary to decide whether the inspection head should be designed to stop the machine in the event of a fault or to prevent further operations being performed on the assembly. In the following studies, it will be assumed that the memory system is incorporated where the inspection head will be designed to activate the memory system rather than to stop the machine.

The general layout of a suitable in-line indexing machine is shown in Fig. 10.6. Note that operations 4 and 6 have been arranged immediately after the

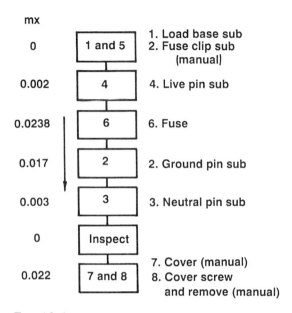

Fig. 10.6 Station layout of in-line indexing machine for assembling power plugs.

first (manual) station to minimize the possibility of the fuse clip becoming displaced during the machine index. When the fuse is in position, the fuse clip is then positively retained. These desirable features provide further restrictions in the order of assembly, and the precedence diagram is modified as shown in Fig. 10.7.

The downtime on an indexing machine is given by the sum of the downtimes on the individual heads due to the feeding of defective parts plus the effective downtime due to the production of unacceptable assemblies.

If, for each machine station, x is the effective proportion of defective to acceptable parts, then mx is the average proportion of defectives that will cause a machine stoppage, and $(1 - m)x$ is the effective average proportion of defectives that will spoil the assembly but not stop the machine. The downtime due to machine stoppages and the final production rate are found as follows.

In the production of N assemblies, the number of machine stoppages is $N\Sigma mx$, where Σmx is the sum of the individual values of mx for the automatic workheads.

If T is the average time to correct a fault and restart the machine, the downtime due to machine stoppages is $NT\Sigma mx$; if t is the machine cycle time, the proportion of downtime D will be given by

$$D = \frac{\Sigma mx}{t/T + \Sigma mx} \tag{10.2}$$

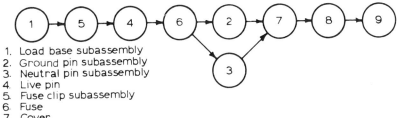

1. Load base subassembly
2. Ground pin subassembly
3. Neutral pin subassembly
4. Live pin
5. Fuse clip subassembly
6. Fuse
7. Cover
8. Cover screw
9. Remove complete assembly

Fig. 10.7 Final precedence diagram for assembly of power plug.

The figures in Table 10.2 are rearranged in Table 10.3 to give the effective quality levels for the various operations. From these figures, it can be seen that the value of Σmx is 0.0678 and, assuming that $t = 7.7$ (the time taken to place the base subassembly on the work fixture and assemble the fuse clip) and that $T = 40$ sec, then,

$$D = \frac{0.0678}{7.7/40 + 0.0678} = 0.26 \ (26\%)$$

During the time the machine is operating, some of the assemblies produced will contain defective parts that did not stop the machine and, assuming that no assembly contains more than one such defective part, the production rate of acceptable assemblies P_a will be given by

$$P_a = \frac{[1 - \Sigma(1 - m)x](1 - D)}{t} \tag{10.3}$$

From Table 10.3, $\Sigma(1 - m)x = 0.0176$ and, therefore, from Eq. (10.3),

$$P_a = \frac{(0.9824)0.74}{7.7} = 0.094 \text{ assemblies/sec } (5.7 \text{ assemblies/min})$$

This corresponds to an overall assembly time of 10.5 sec, and it would be necessary to supplement the machine with some manual assembly in order to meet the required production rate of 7.8 assemblies/min.

Free-Transfer Machine

The layout of a free-transfer machine suitable for assembling the power plug would be the same as that for an indexing machine (Fig. 10.6). It will be assumed in the following that each workstation is capable of accommodating six work carriers, one of which is situated below the workhead, with the remaining five constituting the buffer storage.

Table 10.3 Effective Quality Levels in Assembly of Power Plug

Operation	Automatic station on free-transfer machine	Effective quality level, x	Ratio of defectives causing machine stoppages, m	mx	$(1-m)x$
1. Assemble base subassembly onto work carrier	—	0.001	0	0	0.001
2. Assemble earth pin subassembly into base	4	0.017	1.0	0.017	0
3. Assemble neutral pin subassembly into base	3	0.003	1.0	0.003	0
4. Assemble live pin into base	1	0.002	1.0	0.002	0
5. Assemble fuse clip subassembly into base	—	0	0	0	0
6. Assemble fuse into live pin and fuse clip	2	0.0294	0.813	0.0238	0.0056
7. Assemble cover	5	0.001	0	0	0.001
8. Assemble cover screw	5	0.022	1.0	0.022	0
9. Remove complete assembly	—	0	—	—	—
10. Inspection	—	0.01	0	0	0.01
Totals				0.0678	0.0176

Using a basic cycle time of 7.7 sec and a downtime per fault of 40 sec as before, and the proportions of parts at each station causing a fault (mx) from Table 10.3, a computer simulation was carried out. It was assumed that one technician tended the machine in order to correct faults. The simulation gave an estimated downtime of 14.1% and showed that the technician would be occupied 30% of the time.

The average cycle time for these conditions was 8.96 sec, giving an average assembly rate of 6.7 assemblies/min. Although this figure is higher than that for the indexing machine because of the reduced downtime, the output would again have to be supplemented by manual assembly.

A complete analysis of the automatic assembly of each of the items in the power plug is presented in Table 10.4, where it can be seen that, of the parts that are to be fed automatically, the ground pin sub and the neutral pin sub give a maximum feed rate that is less than the 7.8 parts/min required. For the ground pin sub, this means that an additional 33% can be added to the feeder cost in order to provide the required rate and, for the neutral pin sub, 10% can be added.

Summing the appropriate relative feeder costs gives 6.43 and, multiplying this figure by the basic feeder cost of $7k, the total cost of feeders would be $45k. Similarly, the total relative workhead cost is 7.2 and, for a basic workhead cost of $10k, this gives a total of $72k.

For the indexing machine, assuming a cost per station of $10k for the transfer device and controls and an additional $1k per station for the work carrier, we get $77k for seven stations, giving a total equipment cost of $194k.

In this comparison, we employ the simple payback method for estimating the cost of using equipment and, therefore, using a value of Q_e (the capital equivalent of one assembly worker) of $90k, two-shift working, and an assembly worker rate of 0.83 cent/sec ($30/hr), the rate for the equipment is 0.9 cent/sec.

If the rate for the technician is 1.1 cents/sec, the total for two assembly workers and one technician is 2.8 cents/sec, giving a total rate for the machine and personnel of 3.7 cents/sec. It was estimated earlier that the average assembly time for the indexing machine would be 10.5 sec and, therefore, the assembly cost per assembly would be 38.9 cents. With an efficiency of 80%, this cost increases to 48.6 cents, a figure that can be compared with 45 cents for manual assembly.

For the free-transfer machine, assuming that the cost for each station with five buffers and all controls is $30k and that there are three work carriers per station (costing $1k each), the total for the seven station machine is $348k, including feeders and workheads. This corresponds to a rate of 1.6 cents/sec, giving a total rate for the machine and personnel of 4.4 cents/sec. Finally, for

Table 10.4 Analysis of Power Plug for High-Speed Automatic Assembly

	Handling code	Orienting efficiency	Relative feeder cost	Max feed rate, per min	Insertion code	Relative workhead cost
1 Base sub	64120	0.15	(2)	4.6	(Manual assembly selected)	
2 Fuse clip sub	84044	0.15	(5)	11.3	(Manual assembly selected)	
3 Live pin	71200	0.25	2	10.1	00	1.0
4 Fuse	20000	0.90	1	54.0	31	1.9
5 Ground pin sub	71100	0.15	1.33	5.8	00	1.0
6 Neutral pin sub	71100	0.15	1.1	7.0	00	1.0
7 Inspection	—	—	—	—	—	1.0
8 Cover	64620	0.10	(3)	3.1	(Manual assembly selected)	
9 Cover screw	21000	0.90	1	75.0	48	1.3
Totals			6.43			7.2

NOTE: Codes and data obtained from Ref. 3.

an assembly rate of 6.7 assemblies/min (8.96-sec cycle) and an efficiency of 80%, the total assembly cost is 49.3 cents.

This figure is slightly higher than that for an indexing machine, but this is to be expected because the benefits of the nonsynchronous arrangement become apparent only when machines have relatively large numbers of stations. Also, with both machines, the two assembly workers and the technician are underutilized, leading to a higher cost than that for manual assembly.

One of the principal problems was the manual assembly of the base subassembly, the cover, and the fuse clip subassembly and, ideally, the calculations should be repeated for configurations other than the one considered here. For example, if one assembly worker could be eliminated by feeding and orienting all the parts, the total feeder cost would be $157k, increasing the total equipment costs by $112k. However, an extra station would be required at the beginning of the machine, costing an extra $11k and giving a new total equipment cost for an indexing machine of $317k, an increase of almost 40%. Assuming that the machine could now produce assemblies at the required rate of 7.8/min (i.e., every 7.7 sec), the new assembly cost would be 32.6 cents, including the 80% efficiency factor. However, for comparison purposes, we shall continue with the original configuration, assuming that it is desirable to minimize the investment in equipment.

10.2.6 Robot Assembly of the Power Plug

The use of robots in assembly allows greater flexibility in part-presentation methods. We assume that the base subassembly, cover, and fuse clip subassembly are presented in magazines that have been loaded manually. For each magazine, we make an allowance of $1.5k, and we use the manual handling time for the items, plus 1 sec for insertion into the magazine.

Table 10.5 presents the results of an analysis for a single-station system with one robot arm. A cost of $12k is assumed for the basic work fixture, with an additional $12.5k for a special automatic screwdriver that inserts the cover screw from below. Also, $10k is allowed for the inspection head, which would be attached to the work fixture. A standard SCARA-type robot (estimated cost $80k) can perform all the remaining assembly operations. The total feeder and magazine cost is estimated to be $39.5k, and the assembly worker time for loading magazines is 9.7 sec per assembly. Since these operations are carried out off-line, this time can be multiplied by the assembly worker rate of 0.83 cent/sec, giving a cost of 8.05 cents for magazine loading. Total equipment costs of $154k would give a rate of 0.71 cent/sec. One of the advantages of single-station robot systems is that the downtime due to defective parts forms a relatively small proportion of the total cycle. This means that one technician can tend to several systems. In the present case, the total operation time is 22

Table 10.5 Analysis of Power Plug for Single-Station One-Arm Robot Assembly

	Robot insertion code	Robot cost, k$	Tool cost, k$	Fixture cost, k$	Part presenter cost, k$	Operation time, sec	Time for manual operations, sec
1 Base sub	00	80	—	12.0	1.5	3.0	3.0
2 Fuse clip sub	00	80	—	—	1.5	3.0	3.7
3 Live pin	00	80	—	—	7.0	3.0	—
4 Fuse	30	80	—	—	7.0	3.0	—
5 Ground pin sub	00	80	—	—	7.0	3.0	—
6 Neutral pin sub	00	80	—	—	7.0	3.0	—
7 Inspection	—	—	10	—	—	—	—
8 Cover	02	80	—	—	1.5	3.0	3.0
9 Cover screw	—	—	12.5	—	7.0	1.0	—
Totals			22.5	12.0	39.5	22.0	9.7

NOTE: Codes and data obtained from Ref. 3

sec. The average downtime occurring on each cycle is $T\Sigma mx$ sec or, with Σmx = 0.0678 and T = 40 sec, the downtime per cycle is 2.7 sec, giving a total cycle time of 24.7 sec. Therefore, three systems would be needed to meet the required production volume, which corresponds to one assembly every 7.7 sec while the systems are working. It would seem reasonable to assign one technician to tend all three systems, or an additional rate of 0.37 cent/sec per system.

Thus, the total rate for the equipment and technician is 1.08 cents/sec which, with a cycle time of 24.7 sec, represents a cost of 26.7 cents. Finally, adding the cost of magazine loading (8.05 cents) and allowing for 80% efficiency give an assembly cost of 43.4 cents for a single-station system with one robot arm.

Repeating the calculations for Table 10.6, a system with two robot arms gives a total cycle time of 16.5 sec (including downtime), a total equipment cost of $224k for one machine, and an assembly cost of 41.8 cents. In this calculation, two robot arms were assumed to cost $150k, and the technician was assigned to supervise two systems.

The final system to be considered is the multistation transfer machine with robots. The first step in considering this type of system is to assign tasks to the robots that can be accomplished within the cycle time that would give the production volume required.

In the present example, the required average production time is 7.8 assemblies/min, corresponding to a cycle time of 7.7 sec.

Table 10.7 shows the results of an analysis for a multistation machine where the various operations have been assigned to the workstations. It can be seen that a five-station machine is required. The first three stations have robots carrying out two assembly operations each; the fourth is an inspection station; and, at the fifth, a robot places the cover, and a special screwing device mounted below the station inserts the cover screw.

The cost of each robot is $80k; each station with controls and work carriers costs $33k; the cost of magazines and feeders is $39.5k; the automatic screwdriver costs $12.5k; and the inspection head costs $10k. Summing these figures gives a total equipment cost of $547k. The system cycle time is 6 sec plus downtime, which is found from computer simulation to be 23.4%, with 35 percent of the technician's time spent in correcting faults. This downtime percentage corresponds to an average downtime/cycle of 1.8 sec and, therefore, the average cycle time is 7.8 sec. This figure closely matches that required and corresponds to a production rate of 7.7 assemblies/min.

With a rate for equipment of 2.52 cents/sec, together with the technician rate of 1.1 cents/sec, we get an assembly cost of 28.2 cents. Adding 8.05 cents for magazine loading and allowing for 80% efficiency give a total assembly cost of 45.4 cents.

Table 10.6 Analysis of Power Plug for Single-Station Two-Arm Robot Assembly

	Robot cost, k$	Tool cost, k$	Fixture cost, k$	Part presenter cost, k$	Operation time, sec	Manual operation time, sec
1 Base sub	150	—	12.0	1.5	1.7	3.0
2 Fuse clip sub	150	—	—	1.5	1.7	3.7
3 Fuse	150	—	—	7.0	1.7	—
4 Fuse	150	—	—	7.0	1.7	—
5 Ground pin sub	150	—	—	7.0	1.7	—
6 Neutral pin sub	150	—	—	7.0	1.7	—
7 Inspection	—	10	—	—	—	—
8 Cover	150	—	—	1.5	2.6	3.0
9 Cover Screw	—	12.5	—	7.0	1.0	—
Totals	22.5	12.0	39.5	13.8	9.7

NOTE: Data obtained from Ref. 3

354

Table 10.7 Analysis of Power Plug for Multistation Robot Assembly

	Robot cost, k$	Tool cost, k$	Part presenter cost, k$	Operation time, sec	Work-station no.	Manual operation time, sec
1. Base sub	80	—	1.5	3	1	3.0
2. Fuse clip sub	80	—	1.5	3	1	3.7
3. Live pin	80	—	7.0	3	2	—
4. Fuse	80	—	7.0	3	2	—
5. Ground pin sub	80	—	7.0	3	3	—
6. Neutral pin sub	80	—	7.0	3	3	—
7. Inspection	—	10	—	1	4	—
8. Cover	80	—	1.5	3.1	5	3.0
9. Cover screw	—	12.5	7.0	1	5	—
Totals		22.5	39.5			9.7

NOTE: Data obtained from Ref. 3.

Table 10.8 Comparison of Results of Feasibility Study

| | | Manual assembly | Multistation special-purpose | | Robot systems | | |
			Indexing	Free-transfer	1 arm	2 arm	Multi-station
No. of Systems		6	1	1	3	2	1
No. of Stations:	Manual	1	2	2	0	0	0
	Auto	—	6	6	0	0	1
	Robot	—	—	—	1	1	4
Total system workers	Assembly	6	2	2	0	0	0
	Tech	—	1	1	1	1	1
Total equipment costs, k$		—	194	348	462	448	547
Assembly rate/min		8.4	5.7	6.7	7.3	9.5	7.7
Assembly costs, cents		45.0	48.6	49.3	43.4	41.8	45.4

NOTE: Required rate of assembly while system in operation = 7.8 assembly/min; required cycle time = 7.7 sec

For the example of the power plug and the production conditions selected, all the alternative schemes give assembly costs close to those for manual assembly. The results are summarized in Table 10.8, where it can be seen that the single-station system with two robot arms might yield the best results. However, the capital investment is more than twice that for a special-purpose indexing machine. Also, as was shown, analyses of some other special-purpose machine configurations should be carried out. In fact, it was seen that the use of two assembly workers on the system would not be the most economical arrangement.

REFERENCES

1. Ward, K. A. "Fastening methods in Mechanized Assembly," Paper presented at the Conference on Mechanized Assembly, Salford University, England, July 1966.
2. Prenting, T. O., and Battaglin, R. M., "The Precedence Diagram: A tool for Analysis in Assembly Line Balancing," *Journal of Industrial Engineering*, Vol. 15, July–Aug. 1964 p. 208.
3. Boothroyd, G., and Dewhurst, P., *Product Design for Assembly Handbook*, Boothroyd Dewhurst, Inc., Wakefield, R.I., 1990.

Problems

1. A vibratory-bowl feeder is fitted with springs that are inclined at an angle of 60 deg to the horizontal. The upper portion of the bowl track is horizontal, and the frequency of operation is 60 Hz. The amplitude of vibration is adjusted such that the normal track acceleration has a maximum of 1.0 g for the upper part of the track.

   a. If the coefficient of friction between the parts and the track is 0.5, will forward conveying be achieved by forward sliding only or by a combination of forward and backward sliding?

   b. What will be the maximum parallel track velocity, v_p (mm/sec)?

   c. If the conveying efficiency η is 70%, what will be the mean conveying velocity v_m of the parts? [Note: $\eta = (v_m/v_p) \times 100$.]

2. A standard vibratory-bowl feeder has three leaf springs inclined at 80 deg to the horizontal. These springs are equally spaced around a circle of 225-mm radius and support a bowl that is 600 mm in diameter.

   a. Determine the effective vibration angle for the horizontal upper part of the bowl track.

   b. If the peak-to-peak amplitude of vibration in the line of vibration at the bowl wall is 2.5 mm and the frequency is 60 Hz, determine whether forward conveying will occur and whether this will be by both forward and backward sliding or by forward sliding only. (Assume that the coefficient of friction between the part and the track is 0.5.)

   c. If the feeding motion is 100% efficient, what would be the mean forward conveying velocity of the parts?

3. a. If a vibratory-bowl feeder is driven at a frequency of 60 Hz, with vibration angle of 20 deg, what peak-to-peak horizontal amplitude of vibration (mm) will give a mean conveying velocity of 50 mm/sec on a horizontal track?

   b. For the same conditions as those of part a, what would be the minimum vertical clearance between a part and a wiper blade so that the wiper blade will never reject the parts?

   c. If the leaf springs on the feeder are mounted at a radial position r_2 of 100 mm and the track radius r_1 is 200 mm, what spring angle (measured from the horizontal) will give the vibration angle of 20 deg?

4. A special decal on the side of a vibratory-bowl feeder indicates that the horizontal peak-to-peak amplitude of vibration is 0.25 mm. The angle of the supporting springs is such that the vibration at the bowl wall is inclined at an angle of 20 deg to the horizontal. The coefficient of friction between the parts and the track is 0.2, the frequency of vibration is 60 Hz, and the track is inclined at 5 deg to the horizontal. Determine:

   a. The actual value of A_n/g_n for the inclined track
   b. The value of A_n/g_n for forward sliding to occur during the vibration cycle
   c. The value of A_n/g_n for backward sliding to occur
   d. Whether forward conveying will occur
   e. Whether hopping will occur

5. A vibratory-bowl feeder has a spring angle of 70 deg and is operated at a frequency of 60 Hz. The radius to the springs is 150 mm and to the upper, horizontal part of the track is 200 mm. The amplitude of vibration is set so that the parts traveling around the upper track are on the verge of hopping.

   a. What is the horizontal (parallel) amplitude (a_p) of vibration at the bowl wall? Give the answer in μm.
   b. What is the minimum coefficient of friction between parts and track for forward conveying to occur?
   c. If the coefficient of friction is 0.3, what is the minimum value of the normal amplitude a_n (μm) for forward conveying to occur?

6. Derive an expression for the dimensionless maximum parallel distance jumped J/A_p^0 by a part during each cycle of vibratory feeding (J = parallel distance jumped relative to the track amplitude). Assume that, for maximum distance jumped, $\omega t_2 = \omega t_1 + 2\pi$, where ω is the

frequency in rad/sec, t_1 the time the part leaves the track, and t_2 the time the part lands on the track. Also, assume that the coefficient of friction between the part and the track is such that the part never slides. If the efficiency of conveying is defined by $\eta = v_m(100)/v_{pmax}$, where v_m is the mean conveying velocity and v_{pmax} the maximum parallel track velocity, calculate the maximum efficiency possible under the conditions stated above, with a track angle of zero.

7. A vibratory-bowl feeder operates at a frequency of 60 Hz. The track angle is 5 deg, the vibration angle 40 deg, and the horizontal component of the amplitude of vibration of the track 10 mm. If the coefficient of friction between the parts and the track is 0.2, will the parts be conveyed up the track?

8. The peak-to-peak parallel amplitude of vibration at the wall of a vibratory-bowl feeder is 0.889 mm. The vibration angle is 10 deg, and the frequency of vibration is 60 Hz. Using the figures in the text determine:

    a. The mean conveying velocity
    b. The effective distance the part hops
    c. The height of the part hop
    d. The new conveying velocity if the frequency is reduced to 30 Hz and the amplitude increased threefold

9. Parts in the form of right circular cones having a height H and a base radius R are to be fed and oriented using a vibratory-bowl feeder. The bowl track is designed so that a part will be fed to the orienting devices either standing on its base or lying on its side with the base leading or following. Two active orienting devices are to be used. The first is a step of height h_s, which will cause a part standing on its base to overturn onto its side but will not affect the remaining orientations. The second device consists of a portion of track with a V cross section so that all parts lying on their sides will fall into the V with their bases uppermost.

    Determine the theoretical limits for the dimensionless height of the step (h_s/H) for the situation described, assuming that a very low conveying velocity is to be employed. Assume that the parts do not bounce or slide as they are fed over the step, that no energy is lost as the parts impact with the track, and that $H = 4R$.

10. A rivet feeder (external gate hopper) has a gate (at the lowest point in the feeder body) through which the rivets pass; the gate is $1.4d$ in width, where d is the diameter of the shank of the rivet. The rivets are tumbled within a rotating inner sleeve having slots to accept the rivet shanks that are just wide enough to accept one rivet. The gap between

the inner sleeve and the body of the feeder is equal to 0.6d, and the distance between slots is 4.0d.

    a. Derive an expression, in terms of d, for the critical peripheral velocity of the inner sleeve to prevent jamming.

    b. Estimate the maximum feed rate in rivets/sec if the efficiency with which rivets fall into the slots is 30% and the diameter of the rivet shanks d is 5 mm.

11. A rivet feeder (external gate hopper) has a gate, through which the shanks of the rivets pass; the gate is 1.2d in width, where d is the diameter of the shank of the rivet. The rivets are tumbled within a rotating inner sleeve having slots to accept the rivet shanks that are just wide enough to accept one rivet. The gap between the inner sleeve and the body of the feeder is equal to 0.5d, and the distance between slots is 2.5d. Estimate the maximum feed rate in rivets/sec if the efficiency with which rivets fall into the slots is 30% and the diameter of the rivet shanks d is 5 mm. (Note: The gate is at the bottom of the feeder body.)

12. A rivet feeder (external gate hopper) has a gate, 6.75 mm wide, through which the shanks of the rivets pass. The slots in the rotating inner sleeve are spaced at 15-mm intervals, and the gap between the body of the feeder and the inner sleeve is 2.5 mm. Estimate the maximum feed rate in rivets/min if the diameter of the rivet shank is 5 mm and, on the average, every third slot contains a rivet when the slots pass over the gate, which is located at the bottom of the feeder body.

13. A rivet feeder (external gate hopper) has a gate, 6 mm wide, through which the shanks of the rivets pass, and the diameter of the shank of the rivet is 5 mm. The rivets are tumbled within a rotating inner sleeve having slots to accept the rivet shanks that are just wide enough to accept one rivet. The gap between the inner sleeve and the body of the feeder is 2.5 mm, and the distance between slots is 12.5 mm.

    Estimate the maximum feed rate in rivets/sec if the efficiency with which rivets fall into the slots is 30% (Note: The gate is at the bottom of the feeder body).

14. For an external gate hopper:

    a. determine the maximum value of the gap h_g between the cylinder and the sleeve such that the cylindrical part (diameter D) cannot become jammed between the lower corner of the slot and the inner surface of the sleeve. Assume that the angle of friction for all surfaces is 30 deg.

      b.   For these conditions and for a gate width equal to $3D/2$, find the maximum peripheral velocity v for the inner cylinder in terms of D and g.

15.   A centerboard hopper has a blade length of 262.5 mm and is designed to feed cyclindrical parts end to end. The center of rotation (which is in line with the track) is 250 mm from the lower end of the blade track. The inclination of the track when the blade is in its highest position is 45 deg, and the coefficient of sliding friction between the parts and the track is 0.3. The blade is driven by a cam drive such that the raising of the blade is carried out by a period of uniform acceleration followed by an equal period of deceleration of the same magnitude. The blade then remains stationary to allow parts to slide into the delivery chute and is finally lowered in the same manner as it was raised.

     If the efficiency with which 1-in. (25.4-mm)-long parts are selected by the blade track is 50%, calculate the maximum feed rate possible with these parts (parts/sec) and the corresponding cycle time in seconds.

16.   The blade of a centerboard hopper feeder oscillates vertically as shown in Fig. P.16.

      a.   For the condition in which the blade is full of parts, find the maximum deceleration of the blade so that parts will not lose contact with the track.

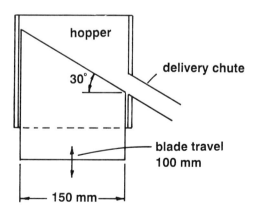

Fig. P.16

b. Assume that the blade is raised 50 mm by accelerating at the value found in problem 16a, decelerates at this value for the next 50 mm, remains stationary to allow all parts to slide off the track, and is finally lowered in the same manner as it was raised. Find the total cycle time in seconds if the coefficient of friction between the parts and the track is 0.3.

c. Find the delivery rate (part/sec), assuming that the efficiency of part selection is 30% and that the parts are 10 mm long.

17. A centrifugal hopper has a diameter of 0.5 m and is designed to feed steel cylindrical parts 20 mm long and 5 mm in diameter. The wall of the hopper is steel and, to increase the feed rate, the base is coated with rubber.

If the coefficient of friction for steel/steel is 0.2 and for steel/rubber is 0.5, estimate:

a. The maximum possible feed rate for the hopper
b. The rotational frequency (rev/sec) that will give this feed rate

18. A centrifugal hopper has a diameter of 0.3 m and is designed to feed steel cyclindrical parts 10 mm long and 4 mm in diameter. The wall of the hopper is steel and, to increase the feed rate, the base is coated with rubber.

If the coefficient of friction for steel/steel is 0.2 and for steel/rubber is 0.5, estimate:

a. The maximum possible feed rate for the hopper (parts/sec)
b. The minimum rotational frequency (rev/sec) that will give this feed rate

19. The part in Fig. P.19 is to be fed and oriented by a double-belt feeder in the orientation depicted. A vision system has been devised that will detect the orientation desired and trigger a mechanism that will reject all other orientations. The belt feeder deposits unoriented parts onto the final delivery belt and then past the vision system at a rate of one part every 1.5 sec.

Estimate the delivery rate of oriented parts, assuming that the coefficient of friction between the part and the guide rail is 0.2; neglect the effects of the cutouts and hole on the probabilities of natural resting aspects and orientations. Also, assume that the belts have "soft" surfaces. Use the graphs in Figs. 3.55 and 3.56, and assume that Fig. 3.56 applies to a part passing along the guide rail of a belt feeder.

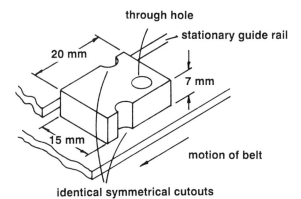

Fig. P.19

20. A V-shaped orienting device for a vibratory-bowl feeder is to be designed to orient truncated cone-shaped parts having a ratio of top diameter to base diameter of 0.8. Good rejection characteristics can be obtained when the half-angle of the cutout is 45 deg. Determine the maximum and minimum distances from the bowl wall to the apex of the cutout (expressed as a ratio of the part base radius) for which all the parts on their tops will be rejected and all the parts on their bases will be accepted. Assume that the parts never leave the track during feeding.

21. A vibratory-bowl feeder is to feed cylindrical cup-shaped parts. The track is to be designed such that only four feeding orientations of the part are possible:

> a: part feeding on its base (27%)
> b_1: part feeding on its side, base leading (35%)
> b_2: part feeding on its side, top leading (35%)
> c: part feeding on its top (3%)

The part has a length-to-diameter ratio of 1.13, and the orienting system comprises three devices:

1. A step that does not affect 49% of the parts in orientation a and that reorients 100% of b_1, 38% of b_2, and 80% of c to orientation a
2. A scallop cutout that rejects all the parts in orientation c

3. A sloped track and ledge that rejects all parts in orientations b_1 and b_2

   If the input rate of parts to the orienting system is 2.5/sec, what will be the output rate of oriented parts from the system?

22. A part that is to be fed and oriented in a vibratory-bowl feeder has five orientations (a, b, c, d, e) on the bowl track. The orienting system consists of three devices. The first is a wiper blade that rejects all the parts in orientation e; the second is a step that reorients 20% of those in orientation a to orientation b, 10% of those in b to orientation c, 50% of those in c to orientation a, and 80% of those in d to orientation a. Write the matrices for each device, and thereby calculate the efficiency of the system if the initial distribution of orientations is as follows:

$$a = 0.2 \qquad b = 0.25 \qquad c = 0.4 \qquad d = 0.1 \qquad e = 0.05$$

What would the efficiency of the system have been if no step had been included?

23. Rectangular prisms are to be fed and oriented in a vibratory-bowl feeder. The two orienting devices employed are a wiper blade and a narrowed track. The probabilities of the important orientations and the dimensions of the prisms are shown in Fig. P.23.

   a. Construct matrices for the wiper blade and narrowed track and, hence, find the system matrix.
   b. By multiplying the system matrix and the orientation matrix, find the efficiency of the orienting system.
   c. Estimate the feed rate of oriented parts (parts/sec) if the conveying velocity is 100 mm/sec and if the parts are contacting each other as they enter the system.

24. The part shown in Fig. P.24 is to be oriented and delivered using a vibratory-bowl feeder. The orienting devices to be used are (1) a wiper blade, (2) a narrowed track, and (3) a scallop cutout.

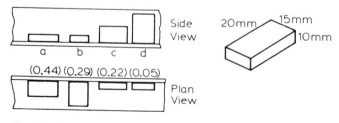

Fig. P.23

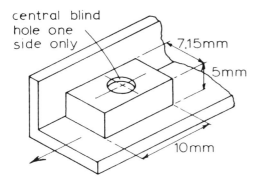

central blind
hole one
side only

7.15mm

5mm

10mm

parts to be delivered in
orientation shown

Fig. P.24

    a.   Estimate the probability of the delivery orientation.

    b.   Estimate the feed rate of oriented parts if parts enter the orienting system at 1/sec on average.

25. A vibratory-bowl feeder orienting system is designed to orient a part that has four orientations on the bowl track. The first device is a step whose performance can be represented by the following matrix:

$$
\begin{array}{c c c c c}
 & a & b & c & d \\
a & \begin{bmatrix} 0.49 & 0.51 & 0 & 0 \\ 1.0 & 0 & 0 & 0 \\ 0.28 & 0.02 & 0.3 & 0.4 \\ 0.80 & 0.1 & 0.1 & 0 \end{bmatrix} \\
b \\
c \\
d
\end{array}
$$

The remaining devices are designed to reject orientations b, c, and d.

    Determine the feed rate of oriented parts if the input rate to the system is 2.5/sec and if the input distribution of orientations is

$$
\begin{array}{c c c c}
a & b & c & d \\
[0.27 & 0.35 & 0.35 & 0.03]
\end{array}
$$

Also, estimate the effect on this performance if a further identical step is added to the beginning of the system.

26. Figure P.26 shows the orientations of a cup-shaped part that is to be fed by a vibratory-bowl feeder. The orientation system consists of a step (for which the performance matrix is given) followed by a scallop

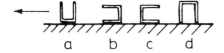

Fig. P.26

that rejects all *d*'s and a sloped track and ledge that rejects all *b*'s and *c*'s. The probabilities of the initial orientations *a*, *b*, *c*, and *d* are 0.45, 0.25, 0.25, and 0.05, respectively.

a. What is the efficiency η of the orienting system?
b. What would the efficiency be if the step was not used?

$$
\begin{array}{c c c c c}
 & a & b & c & d \\
a & \begin{bmatrix} 1 \\ 0.1 \\ 0.8 \\ 0.5 \end{bmatrix} & \begin{matrix} 0 \\ 0.9 \\ 0 \\ 0 \end{matrix} & \begin{matrix} 0 \\ 0 \\ 0.1 \\ 0.5 \end{matrix} & \begin{matrix} 0 \\ 0 \\ 0.1 \\ 0 \end{bmatrix} \\
\end{array}
$$

27. The rectangular prismatic part is to be fed and oriented in a vibratory bowl feeder as shown in Fig. P. 27. Determine the efficiency ($\eta\%$) of this orienting system. Assume a hard track surface, and assume that the coefficient of friction between the bowl wall and that the part is 0.2.

28. The part shown in Fig. P.28 is to be fed and oriented by a vibratory-bowl feeder in the orientation depicted. A television system has been

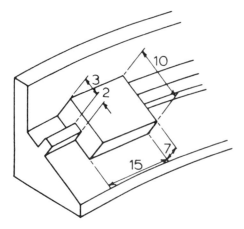

Fig. P.27

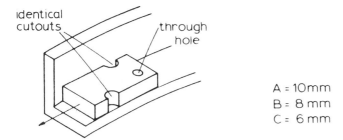

A = 10mm
B = 8 mm
C = 6 mm

Fig. P.28

devised that will detect the orientation desired and trigger a mechanism that will reject all other orientations. What will the feed rate be? Assume that the coefficient of friction between the part and the bowl wall is 0.2. Assume that the conveying velocity of the parts on the bowl track is 20 mm/sec, and neglect the effects of the cutouts and hole on the probabilities of natural resting aspects and orientations. Also, assume a hard surface on the bowl track.

29. Figure P.29 shows a U-shaped metal part that is to be finally oriented by a projection from the wall of a metal vibratory-feeder bowl.

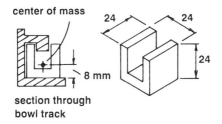

center of mass

8 mm

section through
bowl track

24   24
24

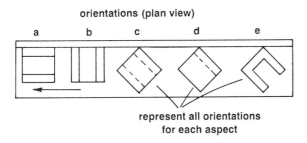

orientations (plan view)

a   b   c   d   e

represent all orientations
for each aspect

Fig. P.29

Find:

a. The probabilities of each of the orientations on the bowl track shown in the figure

b. The efficiency of the orienting system if the projection device shown is preceded by a step in the track that reorients 10% of parts in orientation c, 20% of d, and 20% of e, all equally divided into orientations a and b

c. The feed rate in parts/min if parts enter the system at a rate of 30 parts/min

30. The metal part shown in Fig. P.30 is to be fed and oriented in a vibratory-bowl feeder. The first device is a wiper blade set to reject orientations c–f. The second device is a scallop cutout that rejects all parts in orientation b and 7% of the parts in orientation a.

a. Set up matrices for each device and, by multiplication, obtain the system matrix.

b. From the graphs in the text, determine the complete input distribution matrix (assume $\mu = 0.2$).

c. Find the efficiency for the system.

31. An orienting system for a vibratory-bowl feeder is designed to feed the part shown in Fig. P.31. The important orientations are labeled a, b, c, and d. The first device, a wiper blade, rejects parts in orientation d; then, a narrow track rejects orientation c; and, finally, a scallop cutout rejects orientation b. Assuming metal parts on a metal

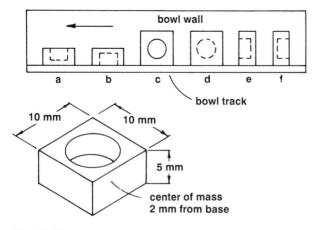

Fig. P.30

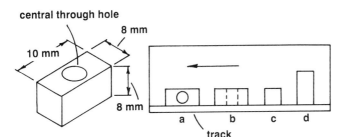

central through hole

10 mm

8 mm

8 mm

a   b   c   d

track

Fig. P.31

track, estimate the feed rate of oriented parts, assuming a mean conveying speed of 20 mm/sec and a coefficient of friction of 0.2. Also, assume that parts enter the orienting system touching each other.

32. A gravity-feed track consists of a straight section of track inclined at an angle of 45 deg, followed by a section of constant radius 10 in. (254 mm), which is followed by a straight horizontal section 5 in. (127 mm) in length. The complete track is arranged in a vertical plane and is to be used in the feeding of parts 0.5 in. (12.7 mm) in length.

   Determine the minimum vertical height (in inches or millimeters) of the top of the column of parts above the inlet to the workhead for feeding to occur if the coefficient of friction between the parts and the track is 0.4.

33. a. A straight gravity-feed track inclined at an angle of 15 deg was found to perform inadequately. A 60-Hz vibrator was attached to the track to improve its performance, and it could achieve a parallel track amplitude of vibration of 0.14 mm. If the coefficient of friction between the parts and the track was 0.3, what would be the velocity of the parts down the track?

   b. Aluminum parts are to be fed on an aluminum, horizontal-delivery gravity-feed track. The track has a straight portion inclined at an angle of 45 deg, followed by a curved portion having a radius of 250 mm and, finally, a horizontal straight portion 50 mm in length. If the parts are 10 mm in length, what will be the minimum delivery time for an escapement placed at the outlet to the track if the height of the column of parts is maintained at 250 mm above the outlet? Assume that the coefficient of friction is 0.3.

34. A straight gravity-feed track is inclined at 15 deg and feeds 10-mm-long parts directly into a slide escapement, which is also inclined at 15

deg. What is the maximum rate at which the slide escapement can operate? Give your answer in cycles/min. Assume that the coefficient of dynamic friction between the parts and the track is 0.26 and that the track is full of parts. It was desired to increase the maximum speed of the escapement by applying a vibration to the track parallel to the track at a frequency of 120 Hz. Using the empirical expression in the text, find the vibration amplitude to give a conveying velocity of 100 mm/sec.

35. The cost of certain screws C is found to be given by

$$C = 10 + 0.01/x \text{ cents}$$

where x is the ratio of faulty screws to acceptable screws. A machine is to be built to assemble 20 such screws (one at each station) into an electrical terminal strip. An assembly worker costing $15/hr will be employed to load the terminal strips (costing 50 cents each), and an engineer costing $20/hr will correct machine faults. The rate for the machine (excluding personnel) is estimated to be $50/hr.

   a. If all faulty parts cause a machine stoppage and if it takes 20 sec to correct each fault, find the *minimum* cost of each complete assembly on an indexing machine working on a 4-sec cycle.
   b. If a robot costing $40/hr could be used to feed, orient, and insert the terminal strips (quality level $x = 0.015$) and if the downtime for each faulty terminal strip was 40 sec, what would the new *minimum* cost be?

36. An indexing assembly machine has 10 stations, and the total cost of running the machine (including operators and overheads) is $0.02/sec. The quality levels x of the parts at each station and the proportions of defective parts that cause a machine stoppage are as shown in Table P. 25. Assume that the time T to correct a machine fault is 30 sec on average and that operator assembly costs are $0.25/assembly. Assume also that the cost of disassembling unacceptable assemblies is $0.80 on average and that the machine has a memory pin system to detect those defective parts that do not stop the machine.

Table P.25 Part Quality Levels

Station	1	2	3	4	5	6	7	8	9	10
x, %	0.12	1.3	1.4	0.5	2.3	0.3	1.6	0.25	0.25	3.6
m	1.0	0.5	1.0	1.0	0	0.5	0.5	1.0	1.0	1.0

At what cycle time t (sec) must the machine be operated so that assembly costs will be the same as those for manual assembly?

37. A 10-station indexing assembly machine is fully automated, and each automatic workhead incorporates a memory pin system such that when a fault is detected, the machine continues to run but no further work is carried out on the assembly containing the fault. If the cycle time of the machine is 5 sec and, at each station, the ratio of unacceptable to acceptable parts is $x\%$, calculate, approximately, the production rate of acceptable assemblies/min when $x = 1.5$

Given that the total running cost of the machine is $1/min, it takes an average of 40 sec to dismantle an unacceptable assembly, operator costs are $0.04/min, and the cost of each part used by the machine is given by $A + B/x$, where $A = B = \$0.004$/part, calculate the number of operators required to dismantle the incomplete assemblies and the total cost of the product. If the parts quality is now changed to the optimum value, determine the optimum and the new total cost of the product.

38. A five-station indexing assembly machine works on a cycle time of 3 sec, and the ratio of unacceptable to acceptable parts used at each station is 0.02. The average time to correct a fault at each station for a machine that stops is 20 sec, and the average time taken to dismantle each part of a faulty assembly on a machine that incorporates a memory pin system is 30 sec. The total running cost of the machine is $0.40/min, and the total cost of each assembly worker used on the machine is W_a cents/min.
Determine:

    a. The production rate and assembly cost per assembly for a machine that stops when a fault occurs
    b. The production rate and assembly cost per assembly for a machine that incorporates the memory system
    c. The value of W_a for which the cost per assembly is the same for both types of machine

Assume that the machine that stops requires only one operator, and calculate the extra assembly workers required on the machine that has a memory pin facility.

39. A five-station indexing assembly machine has a cycle time of 3 sec. The quality levels of the parts supplied to the workheads is 1.1, 2.3, 0.6, 1.5, and 0.2%, respectively. The average time taken to correct a fault and restart the machine is 30 sec. Assuming that every defective part will cause a machine fault, calculate:

a. The percentage downtime for the machine

b. The average production rate (assemblies/min)

If the cost of operating the machine (including overhead) was $20/hr, what maximum amount could be spent over a two-year period to eliminate the defective parts fed to station 2. (Assume 50 weeks working at 40 hr/week.)

40. A five-station indexing assembly machine has a cycle time of 5 sec. The quality level of the parts supplied to the workheads is 1.1, 2.3, 0.6, 1.5, and 0.2%, respectively. The average time taken to correct a fault and restart the machine is 60 sec. Assuming that every defective part will cause a machine fault, estimate:

a. The percentage downtime for the machine

b. The average production rate (acceptable assemblies/min)

Repeat the calculations assuming that only half the defective parts cause a machine fault while the other half spoil the assembly.

41. A rotary indexing assembly machine has six stations. Station 1 involves the manual removal of the completed assembly and the manual loading of the base part of the assembly. The remaining stations are all automatic. These involve the assembly of three pins having quality levels of 1.4% defectives. At the last station, two holding-down screws are inserted, having a quality level of 2% defectives.

If the machine cycle time is 3 sec and the downtime caused by one defective part is 30 sec, determine:

a. The percentage downtime for the machine

b. The average production rate in assemblies/min

If the total machine rate if $16/hr (including operator costs and overheads), would it pay to spend $0.20/100 screws to improve their quality level to 1% defectives?

42. An indexing machine has 12 automatic stations. At stations 1–7, the quality level of the parts is 0.015 defectives; at station 8, the quality is 0.025; and, at the remaining stations, the quality is 0.007. One part is assembled at each station, and each faulty part causes machine downtime of 40 sec. The machine was designed to produce 10 assemblies/min but, unfortunately, the designers neglected to take into account the downtime due to faulty parts.

Estimate:

a. The production rate actually achieved
b. The percentage downtime
c. The machine cycle time needed to achieve the originally planned assembly rate if the downtime due to each fault can be reduced to 20 sec.

43. You have purchased a 3-sec cycle, 20-station indexing assembly machine and, after debugging is complete, the downtime due to faulty parts is 57%. Running costs for the machine are estimated to be $20/hr for the machine and $30/hr for the operator. You are considering converting the machine into two 10-station indexing machines operating at the same speed and have estimated that the cost of conversion will add $5/hr to the running costs. In both situations, the average time to clear a fault and restart the machine is 30 sec. What will be the new downtime and the projected savings or losses due to the conversion? Express the answers in cents per assembly, and assume that the quality levels for all the parts are the same and that all faulty parts cause a machine stoppage.

44. A three-station, free-transfer mechanized assembly machine has a cycle time of 4 sec. At each station, the ratio of defective to acceptable parts is 1:150, and the time taken to correct a fault at any station is 10 sec. Assuming that all stoppages are due to faulty parts, determine the production rate of the machine for equal buffer spaces between each station for 0, 1, 2, 3, 4, and 5 work carriers.

    If the total cost M of running the machine for 1 min is given by

$$M = 0.8 + 0.009b$$

where b is the buffer space between each station, determine the assembly cost per assembly for each of the conditions specified above and, hence, show that an optimum value of b exists for which the assembly costs are minimum.

45. An assembly operation involves 10 operations. No operations can be carried out until operation 1 is complete. Operation 10 cannot be carried out until all other operations are completed. Operation 9 cannot be carried out until operations 7 and 8 are complete. Operations 5, 6, and 7 cannot be carried out until operation 3 has been completed. Operation 8 cannot be carried out until operation 4 is complete. Draw a precedence diagram for this assembly process.

46. Construct a precedence diagram for the 12 parts of the box and lid assembly shown in Fig. P.46. Assume that the first operation is to place the box (part 1) in the work fixture.

47. A feasibility study is to be made for the automatic assembly of three components (Fig. P.47) on an indexing machine that would operate at 1 cycle/sec, with an average downtime due to stoppages of 30 sec. Preliminary studies indicate that the screw has a quality level of 1.5% completely defective. The plastic base has 2.0% defective, but only 0.5% would prevent assembly of the screws; the remainder are cracked or have unacceptable appearance. The metal clips have 1.0% defectives where the screw cannot pass through the hole.

   The estimated total rate for the assembly machine is $23/hr, including one operator and overhead. It is possible to improve the quality levels of the screw and clip to 0.5 and 0.2%, respectively, by means of automatic inspection machines. How much would you be prepared

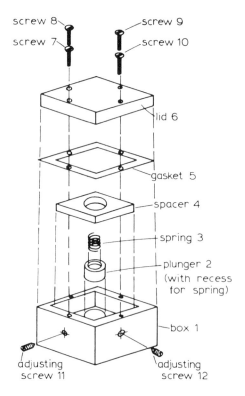

Fig. P.46

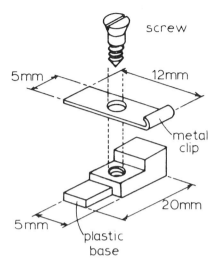

Fig. P.47

to spend on these machines to reduce stoppages on the assembly machine? Neglect the cost of dealing with faulty assemblies produced by the machine, and assume a payback period of 4000 hr and 100% overhead for equipment.

48. Estimate the manual assembly time and the design efficiency for the assembly shown in Fig. P.48. The assembly procedure is as follows:

   1. The retainer bush is placed in the work fixture.
   2. The shaft is inserted into the bush.

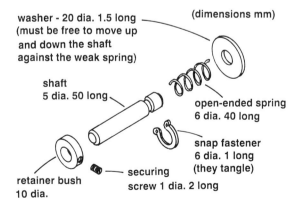

Fig. P.48

3. The securing screw is inserted into the bush and is difficult to align and position.

4. The spring tangles severely and, when separated from the others, is dropped onto the shaft.

5. The washer is inserted onto the shaft and then held down while the snap fastener is positioned using a standard grasping tool.

49. Analyze the gear-box assembly shown in Fig. P.49 for manual assembly.

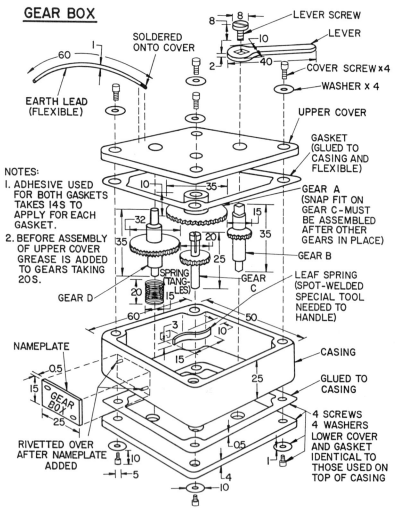

Fig. P.49

Obtain the estimated assembly time, assuming that none of the parts are easy to align and position. Also, obtain the theoretical minimum number of parts, assuming that the springs and the earth lead must be of different materials from the remaining parts. Suggest a redesign for the gear box, assuming that two screws only are needed to secure the upper cover, and estimate the assembly cost for the redesign.

50. Analyze the terminal block shown in Fig. P.50 for manual assembly. Give the estimated assembly time and the design efficiency.

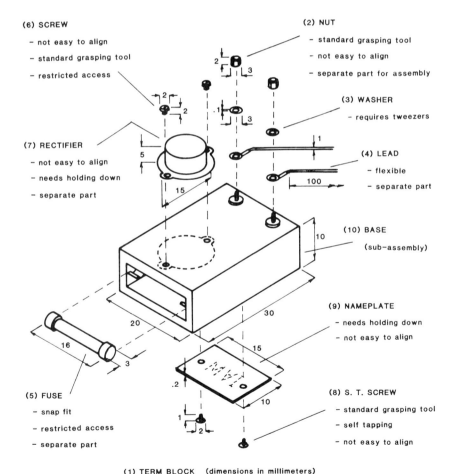

(6) SCREW
- not easy to align
- standard grasping tool
- restricted access

(7) RECTIFIER
- not easy to align
- needs holding down
- separate part

(5) FUSE
- snap fit
- restricted access
- separate part

(2) NUT
- standard grasping tool
- not easy to align
- separate part for assembly

(3) WASHER
- requires tweezers

(4) LEAD
- flexible
- separate part

(10) BASE
(sub-assembly)

(9) NAMEPLATE
- needs holding down
- not easy to align

(8) S. T. SCREW
- standard grasping tool
- self tapping
- not easy to align

(1) TERM BLOCK (dimensions in millimeters)

Fig. P.50

51. Consideration is being given to the purchase of an automatic DIP insertion machine that would be fully occupied inserting 26 fourteen-pin DIPs into a printed circuit board. At present, the DIPs are inserted manually. Estimate the break-even total batch size for the boards if one setup is used for the batch. Include the costs of rework and the cost (150 cents) for a replacement component when rework is carried out. The assembly worker rate is $36/hr, including overheads.

52. A wave-soldered printed circuit board has 25 axial components of 15 types, 32 radial components (2 leads) of 4 types, 2 radial components (3 leads) of 1 type, 14 DIPs (16 pins) of 10 types, and 1 connector (SIP) with 20 pins.

   All the components are manually inserted, but the company is interested in the probable savings if a radial inserter for two lead radials and a DIP inserter is available.

   Estimate the total cost of manual insertion and the savings to be obtained through the use of the auto inserters. Assume a total life volume (total batch size) of 50, two setups, and an assembly worker rate of $45/hr. Neglect the costs of loading the board into the fixture, of wave solder, and of replacement components.

Appendix A

Simple Method for the Determination of the Coefficient of Dynamic Friction

Every student of physics or engineering is aware of the difficulties in obtaining an accurate measurement of the coefficient of dynamic friction between two surfaces. The simple methods normally employed in laboratory demonstrations, methods that everyone is familiar with, are usually unsatisfactory and yield results with large random errors. The improvement in accuracy that can be obtained through repeated measurement is not usually feasible because of the time-consuming adjustment of the inclination of a plane or the adjustment of the load applied to a slider.

An experimental device that will drive a loaded slider across a surface and record the resulting frictional force is necessarily elaborate and expensive and generally justifiable only when research into the frictional behavior of sliding surfaces is undertaken.

The method described here [1] is not intended for this kind of work but is suggested as a replacement for simple undergraduate laboratory experiments on friction or for a rapid measurement of the coefficient of dynamic friction, which is so often desirable in any engineering problem involving relative motion between surfaces.

A.1 THE METHOD

Figure A.1 shows the experimental setup where the slider (1) and the straightedge (2) are placed on a sheet of paper on a horizontal and reasonably flat surface such as a drafting board. A means is provided for constraining the straightedge to move in a straight line over the surface in a direction inclined at an angle θ to the edge itself. This can readily be arranged using a T square and triangle.

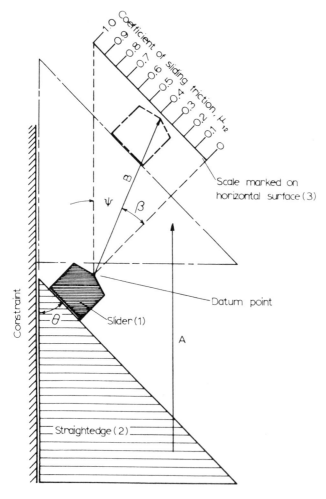

Fig. A.1 Apparatus for determination of coefficient of dynamic friction (from Ref. 1).

As the straightedge is moved at a reasonably uniform speed in the direction of arrow A in Fig. A.1, the slider moves in the direction of arrow B, a direction that (as will be shown later) depends only on the coefficient of sliding friction between the slider and the straightedge and the angle θ. Thus, with the appropriate scale and datum point marked on the horizontal surface, the slider may be moved from the datum point to the scale and the coefficient of friction read off directly.

It will be appreciated that, with this method, repeated measurements can be made rapidly, and a precise determination of the required coefficient of friction can thus be obtained.

The scale marked on the horizontal surface must be appropriate to the chosen angle θ between the straightedge and the direction of its motion. For conditions in which the coefficient of friction lies between zero and unity, a convenient scale is obtained when θ is 45 deg (Fig. A.1). For coefficients of friction greater than unity, a smaller angle of θ must be employed.

A.2 ANALYSIS

The horizontal forces acting on the slider (1) as the straightedge (2) is moved relative to the surface (3) are shown in Fig. A.2. The frictional force between the slider and the surface acts in opposition to the direction of motion of the slider across the surface. This direction is inclined at an angle ψ to the direction of motion of the straightedge; thus, from Fig. A.2:

$$N = \mu_{13}W\sin(\theta + \psi) \tag{A.1}$$

$$F = \mu_{13}W\cos(\theta + \psi) \tag{A.2}$$

where μ_{13} is the coefficient of sliding friction between the slider and the surface and W is the weight of the slider. Now,

$$\mu_{12} = \frac{F}{N} \tag{A.3}$$

and, hence, substitution of Eqs. (A.1) and (A.2) in Eq. (A.3) gives

$$\mu_{12} = \cot(\theta + \psi) \tag{A.4}$$

[note that $(\theta + \psi)$ is the complement of the friction angle β in Fig. A.1] or

$$\tan\psi = \frac{1 - \mu_{12}\tan\theta}{\mu_{12} + \tan\theta} \tag{A.5}$$

It can now be seen that the direction in which the slider moves is independent of both its weight and the coefficient of friction between the slider and the surface.

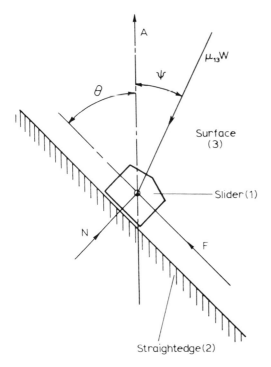

Fig. A.2 Horizontal forces acting on a slider.

Equation (A.5) may be used to construct the required scale for a given value of θ. If θ is 45 deg, a convenient value for θ when $0 \leqslant \mu_{12} \leqslant 1.0$, Eq. (A.5) becomes

$$\tan \psi = \frac{1 - \mu_{12}}{1 + \mu_{12}} \tag{A.6}$$

further, if the scale is arranged parallel to the straightedge, then it is linear with respect to the coefficient of friction (Fig. A.1).

A simpler way of describing the operation of this device is to argue that the slider can move only in the direction of the resultant force applied by the straightedge. This resultant force lies at an angle β (the angle of friction between the slider and the straightedge) to a line drawn normal to the straightedge and, thus, the slider moves as shown in Fig. A.1. Further, since the distance read off the scale is proportional to $\tan \beta$, this scale will be a linear one, giving the coefficient of friction μ_{12}.

An interesting variation of this method is illustrated in Fig. A.3. Here a linear scale in μ_{12} is marked off on the straightedge. The slider is initially set

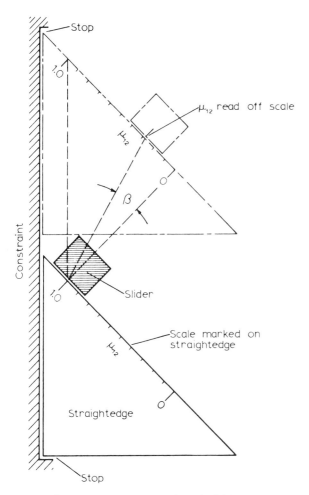

Fig. A.3 Alternative apparatus (from Ref. 1).

at 1.0 on the scale, and the straightedge is moved a distance equal to the length of the scale multiplied by $\sqrt{2}$. The final position of the slider will give the value of the coefficient of friction.

A.3 PRECISION OF THE METHOD

Using a celluloid straightedge and a brass slider, 20 measurements of the coefficient of sliding friction were made. It was found that the mean of the readings was 0.207, with 95% confidence limits of ±0.0043 for the mean. This result indicates the relatively high precision that can be obtained.

A.4 DISCUSSION

It is felt that the simple method for the determination of the coefficient of sliding friction between two surfaces described in this appendix has several advantages over the methods usually employed in undergraduate laboratory demonstrations. These advantages include the following:

1. It requires the minimum of equipment.
2. No delicate adjustments of plane inclination or loading are required.
3. Relatively high precision can be obtained.
4. Repeated readings can be made quickly.
5. Readings are not affected by the static coefficient of friction.
6. Readings are not affected by the small variations in the speed of sliding.
7. It is a direct-reading method.

REFERENCE

1. Boothroyd, G., "Simple Method for the Determination of the Coefficient of Sliding Friction," *Bulletin of Mechanical Engineering Education*, Vol. 9, 1970, p. 219.

Appendix B

Out-of-Phase Vibratory Conveyors

Experimental and theoretical investigations [1] have shown that certain fundamental limitations exist in the performance of conventional vibratory feeders:

1. The conveying velocity of parts up the inclined track is always less than that of parts traveling around the flat bowl base. This means that motion of parts on the track is normally obtained through the pushing action of those circulating around the bowl bottom. With this situation, there is a tendency for parts to jam in the various selecting and orienting devices fitted to the bowl track. Some parts, because of their shape, are difficult to feed under these circumstances. For example, very thin sheet parts are not able to push each other up the track. In this case, the feed rate obtained with a conventional bowl feeder is very low.

Sometimes, it is necessary, as part of the orienting systems, to have a discontinuity in the track. Again, because the parts traveling around the bottom of the bowl cannot push those on the track beyond the discontinuity, the feed rate is generally unsatisfactory.

2. The conveying velocity of parts in a conventional vibratory-bowl feeder is very sensitive to changes in the coefficient of friction between the part and the track, and conveying velocities are very low, with low coefficients of friction.

3.   For high feed rates, it is necessary for the parallel velocity of the track to be high. However, because of the method of driving a conventional vibratory-bowl feeder, an increase in the amplitude of the parallel component of vibration must be accompanied by a corresponding increase in the amplitude of the normal component of vibration. This latter increase is undesirable because, as the normal track acceleration increases above the value that causes the component to hop along the track, the mode of conveying quickly becomes erratic and unstable as a result of the bouncing of parts on impact with the track.

A new drive is described here that is suitable for all types of vibratory conveyors and solves many of the problems associated with conventional designs.

B.1 OUT-OF-PHASE CONVEYING

The new method for driving vibratory feeding devices is based on the idea that the normal and parallel components of motion of the track should have independent amplitude control and should be out of phase. Under these circumstances, the locus of a point on the track becomes elliptical instead of linear.

Theoretical and experimental work was conducted on this new type of drive, [1] and some of its advantages can be demonstrated by means of the results shown in Fig. B.1. In the figure, the product of the mean conveying velocity v_m and the frequency of vibration f is plotted against the phase difference γ between the two components of motion. In the results illustrated, the ratio of the normal a_n and the parallel a_p amplitudes of vibration and the normal track acceleration were both kept constant. The relationships are plotted for three values of the coefficient of friction μ between the part and the track that cover the range likely to be met in practice. It can be seen from the figure that when the phase angle was zero, simulating a conventional feeder, the conveying velocity was very sensitive to changes in μ. Further, for values of μ less than 0.3, the part was moving backward. The results show that if the track parallel motion leads the track normal motion by 65 deg, the conveying velocity becomes uniformly high for all the values considered.

Figure B.2 shows the predicted effect on the mean conveying velocity v_m of changing γ in the relevant range (-90-0 deg) for three values of the amplitude ratio a_n/a_p and when the normal track acceleration A_n is kept constant. In these results, a track angle of 4 deg and a coefficient of friction of 0.2 were chosen because these were considered to represent the most severe conditions likely to be encountered in practice. It is clear from the figure that, for conventional conveying ($\gamma = 0$), as a_p is increased, indicating an increase in the maximum parallel track velocity and a decrease in ψ, the backward conveying

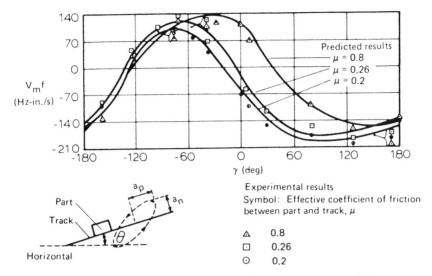

Fig. B.1 Effect of coefficient of friction in out-of-phase conveying. γ is the phase angle, v_m the mean conveying velocity (in./sec), μ the effective coefficient of friction. Track angle θ is 4 deg; vibration frequency f, 21.5 Hz; amplitude ratio a_n/a_p, 0.365; maximum normal track acceleration a_n, 1.2 g.

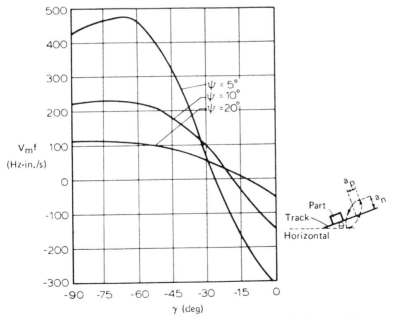

Fig. B.2 Predicted effect of amplitude ratio $(a_n/a_p = \tan \psi)$ in out-of-phase conveying; $\theta = 4$ deg, $\mu = 0.2$, $A_n = 1.2$ g.

velocity of the parts increases. For the optimum phase angle ($\gamma = -65$ deg), however, the forward conveying velocity is increased as a result of an increase in a_p. It is also of interest to note that if the vibration frequency is 25 Hz, a conveying velocity as high as 18 in./sec (400 mm/sec) can be achieved.

These results show that definite advantages are to be gained from operating a vibratory-bowl feeder under the optimum out-of-phase conditions. First, the high conveying velocities attainable are almost independent of the nature of the parts being conveyed. Second, because the feed rate can be controlled by adjusting the parallel component of vibration only, the track normal acceleration may be held constant at a level that does not cause erratic movement of the parts (in the results presented, the normal track acceleration was 1.2 g, which represents stable conveying for most materials). Third, if a_p gradually increased as the part climbs the track, the conveying velocity of the part would gradually increase. This would result in separation of the parts as they climb the track. This situation can be achieved in practice by gradually increasing the track radius, and it would result in more efficient orienting and greater reliability in operation.

B.2 PRACTICAL APPLICATIONS

Figure B.3 shows an exploded view of a vibratory-bowl feeder drive unit designed to operate on the principle outlined above. In this design, motion normal to the track is imparted to the bowl through an intermediate plate supported on the base. Motion parallel to the track is obtained through the springs that support the bowl on the plate. With a suitable controller, the two independent motions will have the required phase difference, and the situation described for out-of-phase conveying can be obtained.

As a practical example of the capabilities of the new design of feeder, an attempt was made to feed thin mica specimens that a manufacturer had previously found almost impossible to feed in a typical conventional vibratory-bowl feeder. With the new feeder, however, it was possible to feed these specimens separately up to the track at conveying velocities of up to 18 in./sec (400 mm/sec) without any erratic motion.

The new type of drive, suitable for all types of vibratory conveyors and called an out-of-phase drive, has many practical advantages. With this type of drive, feeders are quieter; greater flexibility in performance can be achieved; more reliable, yet more sophisticated, orienting devices can be employed; and much higher feed rates can be obtained than with the conventional drive system. (Note: The out-of-phase drive for vibratory conveyors has been patented by the National Research Development Corporation, and inquiries should be directed to the University of Salford Industrial Centre, Salford, Lancashire, England.)

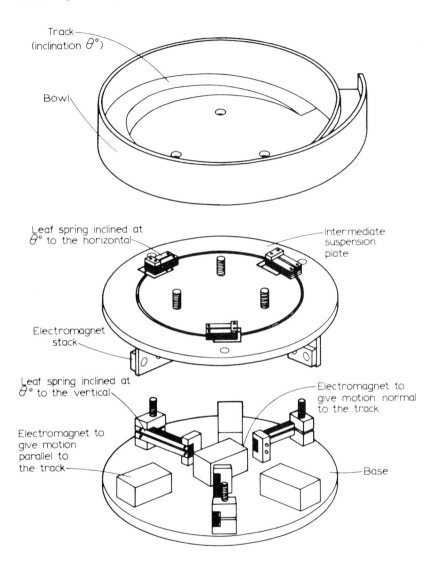

Fig. B.3 Exploded view of a vibratory-bowl feeder drive that has independent motion normal and parallel to the bowl track.

REFERENCE

1. Redford, A.H., "Vibratory Conveyors," Ph.D. thesis, Royal College of Advanced Technology, Salford, England, 1966.

Appendix C

Laboratory Experiments

This appendix gives a complete description of two typical laboratory experiments that may be included in a college or university course on automatic assembly. The first experiment is designed to illustrate certain practical aspects of the performance of a vibratory-bowl feeder; it also indicates how the results of the tests are best presented to gain the maximum information. Clearly, similar experiments could be designed to study the performance of other types of parts-feeding devices employed in automatic assembly. The second experiment illustrates how the coefficient of dynamic friction between small parts and a feed track may be obtained. This information is used in the second part of the experiment to verify the predictions of the theoretical analysis of a horizontal-delivery gravity-feed track.

C.1 PERFORMANCE OF A VIBRATORY-BOWL FEEDER

C.1.1 Objectives

To determine (1) the relationship between vibration amplitude and feed rate for a constant bowl load and (2) the effect of bowl loading on the performance of a vibratory bowl feeder.

C.1.2 Equipment

Vibratory-bowl feeder (10- or 12-in. bowl); 1000 low-carbon steel parts 5/16 in. in diameter and 1 in. long; transducer arranged to measure the vertical component of the bowl vibration amplitude; stopwatch.

C.1.3 Procedure

1. For a range of settings on the bowl amplitude control and with a bowl load of 500 parts, measurements are made of the time taken for a part to travel between two marks scribed on the inside of the bowl. The vertical vibration amplitude of the bowl is also measured and, in these tests, a device is fitted to the top of the bowl track that continuously returns the parts to the bottom of the bowl in order to maintain the bowl load constant.

2. Commencing with the bowl full (1000 parts) and the amplitude control set to give a low feed rate (less than 1 part/sec), the times are measured for successive batches of 100 parts to be delivered. In this test, the parts are allowed to pass down the delivery chute, and the readings are continued until the bowl becomes empty.

C.1.4 Theory

Theoretical and experimental work has shown that the parameters affecting the mean conveying velocity v_m in vibratory conveying are:

1. Maximum track acceleration A (m/s^2)
2. Operating frequency ω (rad/sec)
3. Track angle θ (see Fig. C.1)
4. Vibration angle ψ (see Fig. C.1)
5. Coefficient of friction between component and track μ
6. Acceleration due to gravity g (m/s^2)

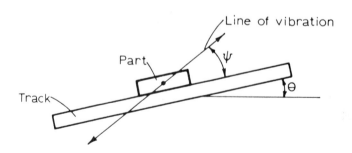

Fig. C.1 Section of track in vibratory-bowl feeder. θ is the track angle and ψ the vibration angle.

Dimensional analysis may now be applied to this problem as follows. Let

$$v_m^{a1} = f[A^{a2}, \omega^{a3}, \theta^{a4}, \psi^{a5}, \mu^{a6}, g^{a7}] \qquad \text{(C.1)}$$

Using the fundamental dimensions of length (L) and time (T), Eq. (C.1) becomes

$$\left[\frac{L}{T}\right]^{a_1} = f\left[\left[\frac{L}{T^2}\right]^{a_2}, \left[\frac{1}{T}\right]^{a_3}, \left[\frac{L}{T^2}\right]^{a_7}\right] \qquad \text{(C.2)}$$

Since the terms θ, ψ, and μ are dimensionless, they have been omitted here to simplify the work. Thus, for Eq. (C.1) to be dimensionally homogeneous,

$$a_1 = a_2 + a_7 \text{ and } -a_1 = -2a_2 - a_3 - 2a_7 \qquad \text{(C.3)}$$

or

$$a_7 = a_1 - a_2 \text{ and } a_3 = -a_1 \qquad \text{(C.4)}$$

Substituting Eqs. (C.4) into Eq. (C.1) yields

$$v_m^{a1} = f\left[A^{a2}, \omega^{-a_1}, \theta^{a4}, \psi^{a5}, \mu^{a6}, g^{a_1-a_2}\right]$$

or, rearranging terms with similar exponents, we obtain

$$\left[\frac{V_m\omega}{g}\right]^{a_1} = f\left[\left[\frac{A}{g}\right]^{a_2}, \theta^{a4}, \psi^{a5}, \mu^{a6}\right] \qquad \text{(C.5)}$$

Thus, for a given bowl and given parts, the dimensionless conveying velocity $v_m\omega/g$ is a function of the dimensionless maximum track acceleration A/g. The theoretical work described in Chapter 3 shows that it is more convenient to employ the component A_n of acceleration normal to the track. It was also shown, however, that conveying is generally achieved by the pushing action of the parts circulating around the flat bowl base. In this case, the effective track angle is zero and, therefore, A, in the analysis above, should be taken as the vertical component A_v of the bowl acceleration.

If it is assumed that the bowl moves with simple harmonic motion, the maximum bowl acceleration may be obtained from measurements of the vertical bowl amplitude and a knowledge of the operating frequency ω.

In the second part of the experiment, the mean feed rate F_0 for each increment in bowl load will be given by

$$F_0 = \frac{100}{t_f} \text{ parts/sec} \qquad \text{(C.6)}$$

where t_f is the time taken (in seconds) to feed 100 parts.

C.1.5 Presentation of Results

Figures C.2 and C.3 show results obtained with a typical commercial bowl feeder. In Fig. C.2, the dimensionless mean conveying velocity $v_m\omega/g$ is plotted against the dimensionless vertical bowl acceleration A_v/g. It can be seen that feeding occurs for all values of A_v/g greater than 0.32 and, from this, it is possible to estimate the coefficient of static friction μ_s between the parts and the track using the analysis described in Chapter 3. Thus,

$$\mu_s = \frac{\cot \psi}{g/A_0 - 1} \tag{C.7}$$

where ψ is the vibration angle, A_0 the minimum vertical acceleration of the bowl for feeding to occur, and g the acceleration due to gravity. In the results presented here, μ_s was estimated to be 0.95 for a mild-steel part in a rubber-coated bowl.

Figure C.3 shows the changes in feed rate F_0 as the bowl gradually empties. It can be seen that, as the bowl load was reduced, the feed rate increased rapidly. Clearly, when the bowl is empty, the feed rate will have fallen to

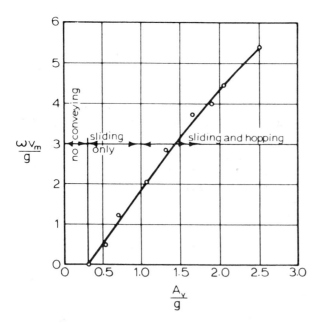

Fig. C.2 Effect of vertical bowl acceleration on conveying velocity for a commercial vibratory-bowl feeder. Mild-steel cylindrical parts 5/16 in. in diameter × 1 in. long; bowl load 500; spring angle 65 deg; vibration frequency 50 Hz.

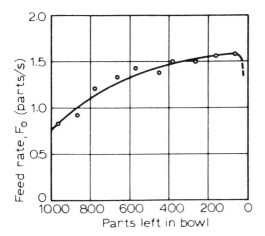

Fig. C.3 Load sensitivity of commercial vibratory-bowl feeder. Mild-steel cylindrical parts 5/16 in. in diameter × 1 in. long; rubber-coated track; spring angle 65 deg; vibration frequency 50 Hz.

zero. For the amplitude setting employed in this test, the bowl could be used to feed the cylindrical parts to a machine or workhead requiring about 40 parts/min. It is clear that, because of the large increase in feed rate as the bowl empties, excessive recirculation of the parts would occur.

C.2 PERFORMANCE OF A HORIZONTAL-DELIVERY GRAVITY-FEED TRACK

C.2.1 Objectives

1. To determine the coefficient of dynamic friction μ_d between the parts and feed track used in the experiment
2. To examine experimentally and theoretically the performance of a horizontal-delivery gravity-feed track

C.2.2 Equipment (Objective 1)

The equipment used in the determination of the coefficient of dynamic friction consisted (Fig. C.4) of a straight track whose angle of inclination θ to the horizontal could be varied between 20 and 60 deg. The equipment is designed to record accurately the time taken for a part to slide from rest a given distance down the track. Near the top of the track, a short peg projecting upward through a hole in the track retains the part until the spring cantilever supporting the peg is deflected by depressing the button. The deflection of the cantil-

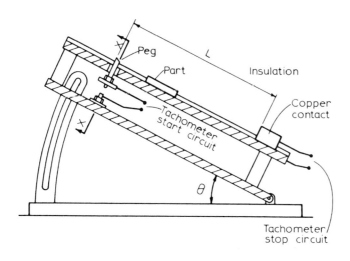

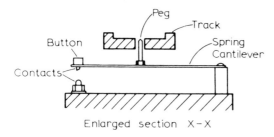

Enlarged section X – X

Fig. C.4 Apparatus used in the determination of the coefficient of dynamic friction.

ever also closes a pair of contacts at the moment the part is released; this initiates the count on a digital clock. A further contact is positioned at the bottom of the track a distance L from the front of the retaining peg. When the part arrives at the bottom of the track, it completes the circuit, which stops the count on the clock.

C.2.3 Theory (Objective 1)

The equation of motion for a part sliding down an inclined track is

$$m_p a = m_p g \sin \theta - \mu_d m_p g \cos \theta$$

or

$$\frac{a}{g} = \sin \theta - \mu_d \cos \theta \tag{C.8}$$

where m_p is the mass of the part, θ the track inclination, a the acceleration of the part, and μ_d the coefficient of dynamic friction between the part and the track.

For a straight inclined track, the acceleration of the part is uniform, and the time t_s taken for the part to slide a distance L is given by

$$t_s^2 = \frac{2L}{a} \tag{C.9}$$

Combining Eqs. (C.8) and (C.9) gives

$$\mu_d = \frac{\sin \theta - 2L/gt_s^2}{\cos \theta} \tag{C.10}$$

C.2.4 Procedure (Objective 1)

In the present experiment, where the part was of mild steel and the track of aluminum alloy, the times were measured for the part to slide a distance of 4 in. (101.6 mm), with the track angle set at 30, 45, and 60 deg. For each condition, 20 readings were taken and averaged, and the 95% confidence limits for each average were computed using the t statistics. The corresponding values of μ_d were computed using Eq. (C.10).

C.2.5 Results (Objective 1)

The mean values of μ_d obtained from each track angle are presented in Table C.1, together with their corresponding 95% confidence limits. Since no significant variation of μ_d with changes in track angle was evident, it was thought reasonable to take the average of all the readings obtained. This yielded a mean value of μ_d of 0.353 with a range, for 95% confidence, of 0.325–0.381.

C.2.6 Equipment (Objective 2)

Figure C.5 shows the design of the experimental horizontal-delivery feed track. A parts-release and timing arrangement similar to that used in the first

Table C.1 Mean Values of μ_d and 95% Confidence Limits

θ, deg	μ_d	95% Confidence limits
30	0.358	±0.028
45	0.354	±0.012
60	0.347	±0.006

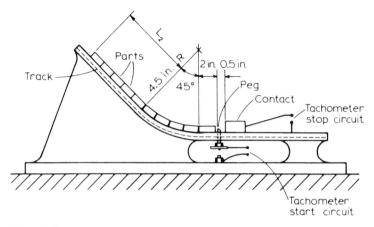

Fig. C.5 Apparatus used to investigate the performance of a gravity-feed track.

part of the experiment is provided. In this case, the column of parts is retained by a peg positioned on the horizontal section 2 in. (50.8 mm) from the beginning of the curved section, and the contact that arrests the column of parts and stops the count is positioned 0.5 in. (12.7 mm) from the front of the peg.

C.2.7 Theory (Objective 2)

The derivation of the theoretical expression for the initial acceleration on release of a column of parts held in a feed track of the present design was developed in Chapter 5. See Eq. (5.10).

Substitution of the values of $L_1 = 2$ in. (50.8 mm), $R = 4.5$ in. (114.3 mm), and $\alpha = 45$ deg for the experimental rig and the mean, upper, and lower values of $\mu_d = 0.353, 0.325,$ and 0.381, respectively, obtained in the first part of the experiment gives the required predicted relationship between a/g and L_2.

In the experiment, the time t_p(sec) was recorded for the column of parts to slide a distance of 0.5 in. (12.7 mm). Since this distance was small compared to the dimensions of the feed track, it could be assumed that the acceleration a of the column of parts was constant. Thus,

$$\frac{a}{g} = \frac{2(0.5)}{g t_p^2} = \frac{2.59 \times 10^{-3}}{t_p^2} \qquad (C.11)$$

C.2.8 Procedure (Objective 2)

The time was recorded for the parts to slide 0.5 in. (12.7 mm) for a range of values of L_2. For each condition, an average of 20 readings was obtained.

C.2.9 Results (Objective 2)

The experimental results are plotted in Fig. C.6, together with the two curves representing the 95% confidence limits. It can be seen that all the experimental results fall within the two theoretical curves representing the 95% confidence limits for the mean value of μ_d.

C.2.10 Conclusions

1. The coefficient of friction between the parts and the track used in the experiment has been successfully determined.

2. The experimental results obtained for the acceleration of a column of parts in a horizontal-delivery gravity-feed track fall within the range, for 95% confidence, of predicted values using the result of the analysis and the values of μ_d obtained in the first part of the experiment. This confirms that the theory, determined for parts of an infinitesimally small length, is valid for parts of finite length, provided that this length is small compared to the dimensions of the track.

3. The performance of parts in a horizontal-delivery gravity-feed track can be accurately estimated using the theoretical equation, provided that the coefficient of dynamic friction between the parts and the track is known.

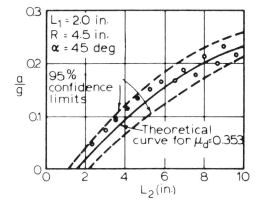

Fig. C.6 Performance of a horizontal-delivery gravity-feed track (steel parts on aluminum track).

Nomenclature

$A =$ Length of longest side of rectangular envelope; maximum track acceleration

$A_f =$ average number of faults requiring rework for automatic insertion of electronic components

$A_i =$ basic cost of one part

$A_n =$ component of maximum track acceleration normal to the track

$A_p =$ component of maximum track acceleration parallel to the track

$B =$ length of intermediate side of rectangular envelope; rate of increase in cost of one part due to quality level

$B_s =$ size of batch to be produced during equipment payback period

$C =$ length of shortest side of rectangular envelope; capital cost of equipment, including overhead

$C_a =$ cost of assembling one assembly

$C_{ai} =$ cost of automatic insertion of electronic component

$C_{ap} =$ programming cost per component for automatic insertion

$C_{as} =$ setup cost per component type for automatic insertion

$C_B =$ cost of transfer device per workstation or buffer space for a free-transfer machine

$C_C =$ cost of work carrier

C_c = cost of replacement component
C_d = dimensionless assembly cost per part
C_e = total equipment cost for an assembly machine
C_F = cost of automatic feeding device and delivery track
C_f = cost of automatic part presentation; cost of feeding one part
C_{gp} = general-purpose equipment cost
C_i = automatic insertion cost for one part; cost of one part
C_M = cost of manually loaded magazine
C_{mi} = cost of manual insertion of one part
C_{mm} = cost of part presentation by manually loaded magazine
C_{op} = assembly operation cost for one electronic component
C_{pr} = average production (assembly) cost for one assembly
C_r = relative feeder cost
C_{ri} = cost of robot insertion of electronic component
C_{rp} = programming cost per component for robot insertion
C_{rs} = setup cost per component type for robot insertion
C_{rw} = average rework cost for electronic component
C_{sp} = special-purpose equipment cost
C_T = cost of transfer device per workstation for an indexing machine
C_t = total cost of each assembly produced, including the cost of parts
$C_{t(min)}$ = minimum total cost of the completed assembly
C_W = cost of workhead
D = diameter of part; diameter of hole; track depth; proportion of downtime on assembly machine
D_h = diameter of rivet head
E = efficiency of feeder; modified efficiency of feeding system; orienting efficiency
E_{ab} = energy barrier in moving a part from orientation a to orientation b
E_{ba} = energy barrier in moving a part from orientation b to orientation a
E_{ma} = manual assembly efficiency
E_o = factory overhead ratio for equipment
F = frictional resistance or force; feed rate
F_m = minimum unrestricted feed rate (that is, the unrestricted feed rate when bowl is full); maximum feed rate for one feeder
F_{max} = maximum feed rate
F_r = required feed rate
FR_s = feed rate per slot for an external gate hopper
H = height the part hops on the track
J = effective distance the part hops on the track
J_0 = J/R
L = length of part; insertion depth

L_1 = length of horizontal track section

L_2 = length of straight inclined track section

M = cost of operating a machine per unit time if only acceptable assemblies are produced

M_f = average number of faults requiring rework for manual insertion of electronic components

M_t = total cost of operating a machine per unit time, including operator's wages, overhead, actual operating costs, machine depreciation, and cost of dealing with unacceptable assemblies

N = normal force; number of vanes; number of parts in hopper; number of assemblies

N_a = number of parts in aspect a

N_b = number of parts in aspect b; number of parts below lower sensor and acceptable level; number of boards per panel in manufacture of PCBs

N_ℓ = number of pins or leads on one electronic component

N_{min} = theoretical minimum number of parts in an assembly

N_p = number of parts delivered per blade, cycle, slot, or revolution; number of parts held in feed track

N_{set} = number of setups for manufacture of PCBs

N_{tech} = minimum number of technician personnel required to correct faults

P = number of parts per vane falling on empty rail

P_a = production rate of acceptable assemblies; probability for aspect a; probability for orientation a

P_b = probability for aspect b; probability for orientation b; equipment payback period (months)

P_e = assembly plant or assembly worker efficiency

P_s = number of shift-years during equipment payback period

P_u = number of unacceptable assemblies produced per unit time

Q_e = equivalent cost of one operator on one shift in terms of capital equipment

R = partition ratio $P_a/(P_a + P_b)$; radius of gravity feed track; radius of base of truncated cone; radius of track

R_e = probability that a part will be rejected

R_f = rate for one automatic feeder; average number of faults requiring rework for robot insertion of electronic components

R_i = rate for one automatic workhead

R_p = number of parts, operations, or components

S_n = number of shifts

T = downtime due to one defective part

T_{rf} = average time to rework one electronic component per lead or post

V_a = annual production volume

V_s = annual production volume per shift

W = weight of part, width of part

W_a = assembly worker rate

W_c = relative workhead cost

W_r = ratio of cost of all personnel to the cost of one manual assembly worker

W_t = total rate for all personnel engaged on an assembly machine

W_{tech} = rate for one technician to correct faults

a = linear acceleration of part

a_n = normal component of amplitude of track vibration

a_0 = amplitude of track vibration

a_p = parallel component of amplitude of track vibration

a_s = center distance between slots of external gate hopper

b = distance from the apex of the V-cutout orienting device to the bowl wall; size of buffer storage between workheads on a free-transfer machine

$b_0 = b/R$

b_t = track width

b_u = largest value of b_0 for which all the unwanted orientations of a truncated cone are rejected

b_w = smallest value of b_0 for which all the wanted orientations of a truncated cone are accepted

c = clearance; dimensionless clearance

d = hopper diameter; diameter of peg; barrel diameter; downtime on workhead due to faulty parts when N assemblies are produced; diameter of screw head

d_g = grip size

d_t = minimum track diameter

f = frequency of vibration

g = acceleration due to gravity

g_n = component of g normal to the track

h = depth of screw head

h_g = gap between cylinder and sleeve in external gate hopper

j = distance as defined in Figs. 3.38 and 3.39

$j_0 = j/R$

ℓ = length of part; length of slot; length of track; width of vanes; distance from the screw head bottom to the center of mass

$\bar{\ell}$ = average part length

ℓ_b = length of barrel

m = proportion of defective parts causing a machine fault, mass of part

m_p = mass of part

m_1 = part mass per unit length

n = reciprocation frequency; rotational frequency; number of parts to be assembled in one assembly; number of parts to be automatically assembled

n_a = number of parts assembled manually per machine cycle

n_c = critical rotational frequency

n_f = average number of parts fed during one workhead cycle with a 100% efficient feeder

n_{max} = maximum reciprocation frequency; maximum rotational frequency

n_r = maximum number of parts assembled at one robot station

n_s = economic number of stations tended by one technician; number of standard deviations

$n_{s\,max}$ = maximum number of stations tended by one technician

n_t = number of robot stations on one machine

$p = [2 + (y/x)^2]^{1/2}$

$q = [1 + (y/x)^2]^{1/2}$

r = radius from center of hopper; blade radius; radius of the top of a truncated cone; radial position

r_b = centerboard hopper swing radius

r_h = radius of hopper hub

$r_0 = r/R$

s = slot width; shank diameter

t = time; cycle time of assembly workhead; average assembly time per part assembled for a programmable workhead or robot; diameter on top of screw head

t_a = manual assembly time for one part (handling and insertion)

t_{at} = assembly machine or station cycle time

t_b = basic insertion time for peg

t_c = time taken by operator to dismantle an unacceptable assembly

t_f = total period of feeder cycle

t_h = handling time for one part

t_i = time for index; manual insertion time

t_m = manual assembly time

t_{ma} = total manual assembly time for one assembly

t_p = time for a part to move one part length; time penalty for insertion of peg through stacked parts

t_{pr} = average production (assembly) time for one acceptable assembly

t_{pw} = additional handling time due to part weight

t_q = required average production time

t_s = time taken for parts to slide from slot of rotary-disk hopper feeder

t_w = workhead cycle time

t_1 = time to lift blade of centerboard hopper

t_2 = dwell time for blade of centerboard hopper

v = peripheral velocity of sleeve in external gate hopper; velocity of part; conveying velocity

v_m = mean conveying velocity

w = width of blade

w_1 = width of chamfer on peg

w_2 = width of chamfer on hole

x = half-width of a square prism; ratio of defective to acceptable parts

x_{opt} = optimum ratio of defective to acceptable parts giving minimum total cost of the completed assembly

y = half-length of a square prism

α = alpha rotational symmetry angle of part; vane angle; inclination angle of gravity-feed track; angle between screw axis and a line normal to the track; cost $(1/D)$

α_1 = arc sin $(1/q)$

α_2 = arc sin $(1/p)$

β = beta rotational symmetry angle of part; arc tan (x/y); friction angle

β_w = wiper blade jamming angle

γ = riser angle; phase angle

η = efficiency of bowl feeder orienting system

θ = track angle; inclination of delivery chute; inclination of elevator hopper; half-angle of the V-cutout orienting device

θ' = inclination of disk in rotary-disk feeder

θ_1 = semiconical angle of chamfer on peg

θ_2 = semiconical angle of chamfer on hole

θ_m = maximum inclination of track

θ_T = critical track angle where screw just touches track cover (or tilt angle)

θ_w = angle between the wiper blade and the bowl wall

λ = hopper inclination for external gate hopper

μ = effective coefficient of friction

μ_b = coefficient of dynamic friction between part and spinning disk

μ_d = coefficient of dynamic friction

μ_{max} = maximum value of coefficient of static friction for sliding to occur

μ_p = coefficient of dynamic friction between part and moving surface

μ_r = coefficient of dynamic friction between part and hook

μ_s = coefficient of static friction

μ_w = coefficient of dynamic friction between part and hopper wall

ϕ = angle of hopper wall

ϕ_g = gate angle for external gate hopper

ψ = vibration angle

ψ_{opt} = optimum vibration angle

ω = angular frequency of vibration; angular velocity

Index